Universitext

Springer
New York
Berlin
Heidelberg
Barcelona
Hong Kong
London
Milan
Paris
Singapore
Tokyo

D1264367

Universitext

Editors (North America): S. Axler, F.W. Gehring, and K.A. Ribet

Aksoy/Khamsi: Nonstandard Methods in Fixed Point Theory
Andersson: Topics in Complex Analysis
Aupetit: A Primer on Spectral Theory
Bachman/Narici/Beckenstein: Fourier and Wavelet Analysis
Bădescu: Algebraic Surfaces
Balakrishnan/Ranganathan: A Textbook of Graph Theory
Balser: Formal Power Series and Linear Systems of Meromorphic Ordinary
  Differential Equations
Bapat: Linear Algebra and Linear Models (2nd ed.)
Berberian: Fundamentals of Real Analysis
Boltyanski/Efremovich: Intuitive Combinatorial Topology. (Shenitzer, trans.)
Booss/Bleecker: Topology and Analysis
Borkar: Probability Theory: An Advanced Course
Böttcher/Silbermann: Introduction to Large Truncated Toeplitz Matrices
Carleson/Gamelin: Complex Dynamics
Cecil: Lie Sphere Geometry: With Applications to Submanifolds
Chae: Lebesgue Integration (2nd ed.)
Charlap: Bieberbach Groups and Flat Manifolds
Chern: Complex Manifolds Without Potential Theory
Cohn: A Classical Invitation to Algebraic Numbers and Class Fields
Curtis: Abstract Linear Algebra
Curtis: Matrix Groups
Debarre: Higher-Dimensional Algebraic Geometry
DiBenedetto: Degenerate Parabolic Equations
Dimca: Singularities and Topology of Hypersurfaces
Edwards: A Formal Background to Mathematics I a/b
Edwards: A Formal Background to Mathematics II a/b
Farenick: Algebras of Linear Transformations
Foulds: Graph Theory Applications
Friedman: Algebraic Surfaces and Holomorphic Vector Bundles
Fuhrmann: A Polynomial Approach to Linear Algebra
Gardiner: A First Course in Group Theory
Gårding/Tambour: Algebra for Computer Science
Goldblatt: Orthogonality and Spacetime Geometry
Gustafson/Rao: Numerical Range: The Field of Values of Linear Operators
  and Matrices
Hahn: Quadratic Algebras, Clifford Algebras, and Arithmetic Witt Groups
Heinonen: Lectures on Analysis on Metric Spaces
Holmgren: A First Course in Discrete Dynamical Systems
Howe/Tan: Non-Abelian Harmonic Analysis: Applications of $SL(2, R)$
Howes: Modern Analysis and Topology
Hsieh/Sibuya: Basic Theory of Ordinary Differential Equations
Humi/Miller: Second Course in Ordinary Differential Equations
Hurwitz/Kritikos: Lectures on Number Theory
Jennings: Modern Geometry with Applications
Jones/Morris/Pearson: Abstract Algebra and Famous Impossibilities
Kac/Cheung: Quantum Calculus

(continued after index)

Victor Kac   Pokman Cheung

Quantum Calculus

Springer

Victor Kac
Department of Mathematics
MIT
Cambridge, MA 02139-2945
USA
kac@math.mit.edu

Pokman Cheung
Department of Mathematics
Stanford University
Stanford, CA 94305-2125
USA
pokman@alum.mit.edu

Mathematics Subject Classification (2000): 81-01, 81R50, 33D05, 05Exx, 68R05

Library of Congress Cataloging-in-Publication Data
Kac, Victor.
    Quantum calculus / Victor Kac, Pokman Cheung.
      p. cm. — (Universitext)
    Includes bibliographical references and index.
    ISBN 0-387-95341-8 (pbk. : alk. paper)
      1. Calculus. I. Kac, Victor G., 1943– II. Title. III. Series.
    QA303 .C537 2001
    515—dc21   2001042965

Printed on acid-free paper.

Production managed by MaryAnn Brickner; manufacturing supervised by Jacqui Ashri.
Photocomposed copy prepared from the authors' T$_E$X files.
Printed and bound by Edwards Brothers, Inc., Ann Arbor, MI.
Printed in the United States of America.

9 8 7 6 5 4 3 2 1

ISBN 0-387-95341-8   SPIN 10847250

Springer-Verlag   New York Berlin Heidelberg
A member of BertelsmannSpringer Science+Business Media GmbH

Contents

Introduction

Consider the following expression:

$$\frac{f(x) - f(x_0)}{x - x_0} .$$

As x approaches x_0, the limit, if it exists, gives the familiar definition of the derivative $\frac{df}{dx}$ of a function $f(x)$ at $x = x_0$. However, if we take $x = qx_0$ or $x = x_0 + h$, where q is a fixed number different from 1, and h a fixed number different from 0, and do not take the limit, we enter the fascinating world of quantum calculus: The corresponding expressions are the definitions of the q-derivative and the h-derivative of $f(x)$. Beginning with these two definitions, we develop in this book two types of quantum calculus, the q-calculus and the h-calculus.

In the course of developing quantum calculus along the traditional lines of ordinary calculus we discover many important notions and results in combinatorics, number theory, and other fields of mathematics.

For example, the q-derivative of x^n is $[n]x^{n-1}$, where

$$[n] = \frac{q^n - 1}{q - 1}$$

is the q-analogue of n (in the sense that n is the limit of $[n]$ as $q \to 1$). Next, in the search of the q-analogue of the binomial, that is a function, $(x - a)_q^n$ that "behaves" with respect to the q-derivative in the same way as $(x - a)^n$ "behaves" with respect to the ordinary derivative, we discover the function

$$(x - a)_q^n = (x - a)(x - qa) \cdots \left(x - q^{n-1}a\right) .$$

The quantity $(1 - a)_q^n$ plays in combinatorics the most fundamental role, and we find it unfortunate that the commonly used notation $(a; q)_n$ for this quantity is so nonsuggestive.

Having the q-binomial, we go on to establish a q-analogue of Taylor's formula. Remarkably, the q-Taylor formula encompasses many results of eighteenth and nineteenth century mathematics: Euler's identities for q-exponential functions, Gauss's q-binomial formula, and Heine's formula for a q-hypergeometric function.

Of course, Gauss's formula

$$(x + a)_q^n = \sum_{j=0}^{n} q^{j(j-1)/2} \begin{bmatrix} n \\ j \end{bmatrix} a^j x^{n-j}$$

is the source of the all important q-binomial coefficients

$$\begin{bmatrix} n \\ j \end{bmatrix} = \frac{[n]!}{[j]![n-j]!}, \quad \text{where } [k]! = [1][2] \cdots [k].$$

We study these coefficients in some detail; in particular, we interpret them in terms of geometry over finite fields.

Euler's identities lead to the celebrated Jacobi triple-product identity, and Heine's formula leads to the remarkable Ramanujan product formula.

Having established all these formulas, we go on to harvest the whole array of applications, rediscovering some of the famous results of eighteenth and nineteenth century mathematics: Euler's recurrent formula for the classical partition function, Gauss's formula for the number of sums of two squares, Jacobi's formula for the number of sums of four squares, etc. The special cases of the last two results are, of course, Fermat's theorem that an odd prime p can be represented as a sum of two squares of integers if and only if $p - 1$ is divisible by 4, and Lagrange's theorem that any positive integer is a sum of four squares of integers.

Returning to q-calculus, as in the ordinary calculus, after studying the properties of the q-derivative we go on to study the q-antiderivative and the definite q-integral. The latter was introduced by F.H. Jackson in the beginning of the twentieth century: He was the first to develop q-calculus in a systematic way.

We conclude our treatment of q-calculus with a study of q-analogues of classical Euler's gamma and beta functions.

In spite of its apparent similarity to q-calculus, the h-calculus is rather different. It is really the calculus of finite differences, but a more systematic analogy with classical calculus makes it more transparent. For example, the h-Taylor formula is nothing else but Newton's interpolation formula, and h-integration by parts is simply the Abel transform. The definite h-integral is a Riemann sum, so that the fundamental theorem of h-calculus allows one to evaluate finite sums.

We do this for sums of nth powers, using the $(h = 1)$-integral of x^n. This leads us naturally to Bernoulli numbers and Bernoulli polynomials. Closely related is the Euler–Maclaurin formula, discussed at the end of the book.

The book assumes only some knowledge of first-year calculus and linear algebra and is addressed mainly to undergraduate students (the second author was an undergraduate during the preparation of the book).

This book is based on lectures and seminars given by the first author at MIT: A part of the lecture course on quantum groups in the fall of 1993, a seminar in analysis for majors in the fall of 1996, and the freshman seminar on quantum calculus in the spring of 2000, in which the second author was the most active participant. We are grateful to the Undergraduate Research Opportunities Program at MIT for their support. We are also grateful to Dan Stroock for very useful suggestions.

In our presentation of the Euler–Maclaurin formula we used unpublished lecture notes by Haynes Miller. We wish to thank him for giving us these notes. Other sources that have been used are quoted at the end of the book.

1
q-Derivative and h-Derivative

As has been mentioned in the introduction, we shall develop two types of quantum calculus, the q-calculus and the h-calculus. We begin with the notion of a quantum differential.

Definition. Consider an arbitrary function $f(x)$. Its q-*differential* is

$$d_q f(x) = f(qx) - f(x), \qquad (1.1)$$

and its h-*differential* is

$$d_h f(x) = f(x+h) - f(x). \qquad (1.2)$$

Note that in particular, $d_q x = (q-1)x$ and $d_h x = h$. An interesting difference of the quantum differentials from the ordinary ones is the lack of symmetry in the differential of the product of two functions. Since

$$
\begin{aligned}
d_q\big(f(x)g(x)\big) &= f(qx)g(qx) - f(x)g(x) \\
&= f(qx)g(qx) - f(qx)g(x) + f(qx)g(x) - f(x)g(x),
\end{aligned}
$$

we have

$$d_q\big(f(x)g(x)\big) = f(qx)d_q g(x) + g(x)d_q f(x), \qquad (1.3)$$

and similarly,

$$d_h\big(f(x)g(x)\big) = f(x+h)d_h g(x) + g(x)d_h f(x). \qquad (1.4)$$

With the two quantum differentials we can then define the corresponding quantum derivatives.

Definition. The following two expressions,

$$D_q f(x) \;=\; \frac{d_q f(x)}{d_q x} = \frac{f(qx) - f(x)}{(q-1)x}, \tag{1.5}$$

$$D_h f(x) \;=\; \frac{d_h f(x)}{d_h x} = \frac{f(x+h) - f(x)}{h}, \tag{1.6}$$

are called the *q-derivative* and *h-derivative*, respectively, of the function $f(x)$.

Note that

$$\lim_{q \to 1} D_q f(x) = \lim_{h \to 0} D_h f(x) = \frac{df(x)}{dx}$$

if $f(x)$ is differentiable. The Leibniz notation $\frac{df(x)}{dx}$, a ratio of two "infinitesimals," is rather confusing, since the notion of the differential $df(x)$ requires an elaborate explanation. In contrast, the notions of q- and h-differentials are obvious, and the q- and h-derivatives are plain ratios.

It is clear that as with the ordinary derivative, the action of taking the q- or h-derivative of a function is a linear operator. In other words, D_q and D_h have the property that for any constants a and b,

$$D_q\big(af(x) + bg(x)\big) \;=\; aD_q f(x) + bD_q g(x),$$
$$D_h\big(af(x) + bg(x)\big) \;=\; aD_h f(x) + bD_h g(x).$$

Example. Compute the q-derivative and h-derivative of $f(x) = x^n$, where n is a positive integer. By definition,

$$D_q x^n = \frac{(qx)^n - x^n}{(q-1)x} = \frac{q^n - 1}{q - 1} x^{n-1}, \tag{1.7}$$

and

$$D_h x^n = \frac{(x+h)^n - x^n}{h} = nx^{n-1} + \frac{n(n-1)}{2} x^{n-2} h + \cdots + h^{n-1}. \tag{1.8}$$

Since the fraction $(q^n - 1)/(q - 1)$ appears quite frequently, let us introduce the following notation,

$$[n] = \frac{q^n - 1}{q - 1} = q^{n-1} + \cdots + 1, \tag{1.9}$$

for any positive integer n. This is called the *q-analogue* of n. Then (1.7) becomes

$$D_q x^n = [n] x^{n-1}, \tag{1.10}$$

which resembles the ordinary derivative of x^n. As $q \to 1$, we have $[n] = q^{n-1} + \cdots + 1 \to 1 + 1 + \cdots + 1 = n$. As we shall see time and again, $[n]$ plays the same role in q-calculus as the integer n does in ordinary calculus.

On the other hand, the expression of $D_h x^n$ is more complicated. It is fair to say that x^n is a good function in q-calculus but a bad one in h-

calculus. For the time being, we will focus on q-calculus. The h-calculus will be discussed in the last chapters of the book.

Let us compute the q-derivative of the product and the quotient of $f(x)$ and $g(x)$. From (1.3) we have

$$D_q\big(f(x)g(x)\big) = \frac{d_q\big(f(x)g(x)\big)}{(q-1)x} = \frac{f(qx)d_qg(x) + g(x)d_qf(x)}{(q-1)x},$$

and hence,

$$D_q\big(f(x)g(x)\big) = f(qx)D_qg(x) + g(x)D_qf(x). \tag{1.11}$$

By symmetry, we can interchange f and g, and obtain

$$D_q\big(f(x)g(x)\big) = f(x)D_qg(x) + g(qx)D_qf(x), \tag{1.12}$$

which is equivalent to (1.11).

If we apply (1.11) to differentiate

$$g(x) \cdot \frac{f(x)}{g(x)} = f(x),$$

we obtain

$$g(qx)D_q\left(\frac{f(x)}{g(x)}\right) + \frac{f(x)}{g(x)}D_qg(x) = D_qf(x),$$

and thus,

$$D_q\left(\frac{f(x)}{g(x)}\right) = \frac{g(x)D_qf(x) - f(x)D_qg(x)}{g(x)g(qx)}. \tag{1.13}$$

However, if we use (1.12), we get

$$g(x)D_q\left(\frac{f(x)}{g(x)}\right) + \frac{f(qx)}{g(qx)}D_qg(x) = D_qf(x),$$

and thus,

$$D_q\left(\frac{f(x)}{g(x)}\right) = \frac{g(qx)D_qf(x) - f(qx)D_qg(x)}{g(x)g(qx)}. \tag{1.14}$$

The formulas (1.13) and (1.14) are both valid, but one may be more useful than the other under particular circumstances.

After deriving the product rule and quotient rule of q-differentiation, one may then wonder about a quantum version of the chain rule. However, *there doesn't exist a general chain rule for q-derivatives.* An exception is the differentiation of a function of the form $f(u(x))$, where $u = u(x) = \alpha x^\beta$

with α, β being constants. To see how chain rule applies, consider

$$
\begin{aligned}
D_q\left[f(u(x))\right] &= D_q\left[f\left(\alpha x^\beta\right)\right] = \frac{f(\alpha q^\beta x^\beta) - f\left(\alpha x^\beta\right)}{qx - x} \\
&= \frac{f\left(\alpha q^\beta x^\beta\right) - f\left(\alpha x^\beta\right)}{\alpha q^\beta x^\beta - \alpha x^\beta} \cdot \frac{\alpha q^\beta x^\beta - \alpha x^\beta}{qx - x} \\
&= \frac{f\left(q^\beta u\right) - f(u)}{q^\beta u - u} \cdot \frac{u(qx) - u(x)}{qx - x},
\end{aligned}
$$

and hence,

$$
D_q f(u(x)) = \left(D_{q^\beta} f\right)(u(x)) \cdot D_q u(x). \tag{1.15}
$$

On the other hand, if for instance $u(x) = x + x^2$ or $u(x) = \sin x$, the quantity $u(qx)$ cannot be expressed in terms of u in a simple manner, and thus it is impossible to have a general chain rule.

We end this section with a discussion of why the letters h and q are used as the parameters. The letter q has several meanings:

- the first letter of "quantum,"

- the letter commonly used to denote the number of elements in a finite field,

- the indeterminate of power series expansions.

The letter h is used as a reminder of Planck's constant, which is the most important fundamental physical constant in quantum mechanics (physics of the microscopic world). One gets the "classical" limit as $q \to 1$ or $h \to 0$, and the two quantum parameters are usually related by $q = e^h$.

2
Generalized Taylor's Formula for Polynomials

In the ordinary calculus, a function, $f(x)$ that possesses derivatives of all orders is *analytic* at $x = a$ if it can be expressed as a power series about $x = a$. Taylor's theorem tells us the power series is

$$f(x) = \sum_{n=0}^{\infty} f^{(n)}(a) \frac{(x-a)^n}{n!}. \tag{2.1}$$

The Taylor expansion of an analytic function often allows us to extend the definition of the function to a larger and more interesting domain. For example, we can use the Taylor expansion of e^x to define the exponentials of complex numbers and square matrices. We would also like to formulate a q-analogue of Taylor's formula. But before doing so, let us first consider a more general situation.

Theorem 2.1. *Let a be a number, D be a linear operator on the space of polynomials, and $\{P_0(x), P_1(x), P_2(x), \ldots\}$ be a sequence of polynomials satisfying three conditions:*

(a) $P_0(a) = 1$ and $P_n(a) = 0$ for any $n \geq 1$;
(b) $\deg P_n = n$;
(c) $DP_n(x) = P_{n-1}(x)$ for any $n \geq 1$, and $D(1) = 0$.
Then, for any polynomial $f(x)$ of degree N, one has the following generalized Taylor formula:

$$f(x) = \sum_{n=0}^{N} (D^n f)(a) P_n(x). \tag{2.2}$$

Proof. Let V be the space of polynomials of degree not larger than N, so that $\dim V = N + 1$. The polynomials $\{P_0(x), P_1(x), \ldots, P_N(x)\}$ are linearly independent because, by condition (b), their degrees are strictly increasing. Hence they constitute a basis for V; i.e., any polynomial $f(x) \in V$ may be expressed as

$$f(x) = \sum_{k=0}^{N} c_k P_k(x) \tag{2.3}$$

for some unique constants c_k. Putting $x = a$ and using condition (a), one gets $c_0 = f(a)$. Then, apply the linear operator D n times to both sides of the above equation, where $1 \le n \le N$. Using (b) and (c), we get

$$(D^n f)(x) = \sum_{k=n}^{N} c_k D^n P_k(x) = \sum_{k=n}^{N} c_k P_{k-n}(x) \,.$$

Again, putting $x = a$ and using (a), we get

$$c_n = (D^n f)(a), \quad 0 \le n \le N,$$

and (2.3) becomes (2.2). □

Example. If

$$D = \frac{d}{dx}, \quad P_n(x) = \frac{(x-a)^n}{n!},$$

then all the three conditions are satisfied, and the theorem gives the Taylor expansion about a of a polynomial.

It is easy to see that given D, the sequence of polynomials satisfying conditions (a), (b), and (c) of Theorem 2.1, if it exists, is uniquely determined. Moreover, if D is a linear operator that maps the space of polynomials of degree n onto the space of polynomials of degree $n - 1$, such a sequence always exists.

3

q-Analogue of $(x - a)^n$, n an Integer, and q-Derivatives of Binomials

As remarked in Chapter 1, D_q is a linear operator on the space of polynomials. We shall try to apply Theorem 2.1 to $D \equiv D_q$. We shall need for that the following q-analogue of $n!$:

$$[n]! = \begin{cases} 1 & \text{if } n = 0, \\ [n] \times [n-1] \times \cdots \times [1] & \text{if } n = 1, 2, \ldots. \end{cases} \tag{3.1}$$

Now let us construct the sequence of polynomials $\{P_0(x), P_1(x), P_2(x), \ldots\}$ satisfying the three conditions of Theorem 2.1 with respect to $D \equiv D_q$. If $a = 0$, we can choose

$$P_n(x) = \frac{x^n}{[n]!}, \tag{3.2}$$

because (a) $P_0(0) = 1$, $P_n(0) = 0$ for $n \geq 1$, (b) $\deg P_n = n$, and (c) using (1.10), for $n \geq 1$,

$$D_q P_n(x) = \frac{D_q x^n}{[n]!} = \frac{[n] x^{n-1}}{[n]!} = \frac{x^{n-1}}{[n-1]!} = P_{n-1}(x).$$

If $a \neq 0$, $P_n(x)$ is not simply $(x-a)^n/[n]!$; for example, $D_q(x-a)^2/[2]! \neq (x-a)$. Let us find the first few $P_n(x)$ and try to deduce a general formula. We have

$$P_0(x) = 1.$$

In order that $D_q P_1(x) = 1$ and $P_1(a) = 0$, we must have

$$P_1(x) = x - a.$$

In order that $D_q P_2(x) = x - a$ and $P_2(a) = 0$, we must have

$$P_2(x) = \frac{x^2}{[2]} - ax - \frac{a^2}{[2]} + a^2 = \frac{(x-a)(x-qa)}{[2]}.$$

Similarly,

$$P_3(x) = \frac{(x-a)(x-qa)(x-q^2a)}{[2][3]},$$

and so on. A logical guess would be

$$P_n(x) = \frac{(x-a)(x-qa)\cdots(x-q^{n-1}a)}{[n]!}, \tag{3.3}$$

which agrees with (3.2) when $a = 0$. Before verifying the validity of condition (c) for Theorem 2.1, let us introduce some notation.

Definition. The q-analogue of $(x-a)^n$ is the polynomial

$$(x-a)_q^n = \begin{cases} 1 & \text{if } n = 0, \\ (x-a)(x-qa)\cdots(x-q^{n-1}a) & \text{if } n \geq 1. \end{cases} \tag{3.4}$$

Proposition 3.1. *For* $n \geq 1$,

$$D_q(x-a)_q^n = [n](x-a)_q^{n-1}. \tag{3.5}$$

Proof. The formula is obviously true when $n = 1$. Let us assume $D_q(x-a)_q^k = [k](x-a)_q^{k-1}$ for some integer k. According to the definition, $(x-a)_q^{k+1} = (x-a)_q^k(x-q^ka)$. Using the product rule (1.12),

$$\begin{aligned} D_q(x-a)_q^{k+1} &= (x-a)_q^k + (qx - q^ka)D_q(x-a)_q^k \\ &= (x-a)_q^k + q(x - q^{k-1}a) \cdot [k](x-a)_q^{k-1} \\ &= (1+q[k])(x-a)_q^k = [k+1](x-a)_q^k. \end{aligned}$$

Hence, the proposition is proved by induction on k.   □

Thus, $D_q P_n = P_{n-1}$ is an immediate result of the above proposition. Now let us explore some other properties of the polynomial $(x-a)_q^n$.

In general, $(x-a)_q^{m+n} \neq (x-a)_q^m(x-a)_q^n$. Instead,

$$\begin{aligned} (x-a)_q^{m+n} &= (x-a)(x-qa)\cdots(x-q^{m-1}a)(x-q^ma)(x-q^{m+1}a) \\ &\quad \times \cdots (x-q^{m+n-1}a) \\ &= \left((x-a)(x-qa)\cdots(x-q^{m-1}a)\right) \\ &\quad \times \left((x-q^ma)(x-q(q^ma))\cdots(x-q^{n-1}(q^ma))\right), \end{aligned}$$

which gives

$$(x-a)_q^{m+n} = (x-a)_q^m(x-q^ma)_q^n. \tag{3.6}$$

Substituting m by $-n$, we can thus extend the definition in (3.4) to all integers by defining

$$(x-a)_q^{-n} = \frac{1}{(x-q^{-n}a)_q^n},\tag{3.7}$$

for any positive integer n. The following two propositions show that this indeed gives a good extension.

Proposition 3.2. *For any two integers m and n, (3.6) is true.*

Proof. The case where $m > 0$ and $n > 0$ has already been proved, and the case where one of m and n is zero is easy. Let us first consider $m = -m' < 0$ and $n > 0$. Then,

$$
\begin{aligned}
(x-a)_q^m(x-q^m a)_q^n &= (x-a)_q^{-m'}(x-q^{-m'}a)_q^n \\
\text{by (3.7)} \quad &= \frac{(x-q^{-m'}a)_q^n}{(x-q^{-m'}a)_q^{m'}} \\
\text{by (3.6)} \quad &= \begin{cases} (x-q^{m'}(q^{-m'}a))_q^{n-m'} & n \geq m' \\ \frac{1}{(x-q^n(q^{-m'}a))_q^{m'-n}} & n < m' \end{cases} \\
\text{by (3.7)} \quad &= (x-a)_q^{n-m'} = (x-a)_q^{n+m} .
\end{aligned}
$$

If $m \geq 0$ and $n = -n' < 0$, then

$$
\begin{aligned}
(x-a)_q^m(x-q^m a)_q^n &= (x-a)_q^m(x-q^m a)_q^{-n'} \\
\text{by (3.7)} \quad &= \frac{(x-a)_q^m}{(x-q^{m-n'}a)_q^{n'}} \\
\text{by (3.6)} \quad &= \begin{cases} \frac{(x-a)_q^{m-n'}(x-q^{m-n'}a)_q^{n'}}{(x-q^{m-n'}a)_q^{n'}} & m \geq n \\ \frac{(x-a)_q^m}{(x-q^{m-n'}a)_q^{n'-m}\left(x-q^{n'-m}(q^{m-n'}a)\right)_q^m} & m < n \end{cases} \\
&= \begin{cases} (x-a)_q^{m-n'} & m \geq n \\ \frac{1}{(x-q^{m-n'}a)_q^{n'-m}} & m < n \end{cases} \\
&= (x-a)_q^{m-n'} = (x-a)_q^{m+n} .
\end{aligned}
$$

Lastly, if $m = -m' < 0$ and $n = -n' < 0$,

$$
\begin{aligned}
(x-a)_q^m (x - q^m a)_q^n &= (x-a)_q^{-m'} (x - q^{-m'} a)_q^{-n'} \\
&= \frac{1}{(x - q^{-m'} a)_q^{m'} (x - q^{-n'-m'} a)_q^{n'}} \\
&= \frac{1}{(x - q^{-n'-m'} a)_q^{n'} \left(x - q^{n'} (q^{-m'-n'} a) \right)_q^{m'}} \\
&= \frac{1}{(x - q^{-n'-m'} a)_q^{n'+m'}} \\
&= (x-a)_q^{-m'-n'} = (x-a)_q^{m+n}.
\end{aligned}
$$

Therefore, (3.6) is true for any integers m and n.   □

We would like to see that Proposition 3.1 is true for any integer n as well. But before proving this, we have to extend our definition of $[n]$ in (1.9).

Definition. For any number α,

$$[\alpha] = \frac{1 - q^\alpha}{1 - q}. \tag{3.8}$$

Proposition 3.3. *For any integer n,*

$$D_q(x-a)_q^n = [n](x-a)_q^{n-1}.$$

Proof. Note that $[0] = 0$, so (3.8) is true for $n = 0$. If $n = -n' < 0$, using (1.13) and (3.7) we have

$$
\begin{aligned}
D_q(x-a)_q^n &= D_q \left(\frac{1}{(x - q^{-n'} a)_q^{n'}} \right) \\
&= -\frac{D_q(x - q^{-n'} a)_q^{n'}}{(x - q^{-n'} a)_q^{n'} (qx - q^{-n'} a)_q^{n'}} \\
&= -\frac{[n'](x - q^{-n'} a)_q^{n'-1}}{q^{n'}(x - q^{-n'} a)_q^{n'} (x - q^{-n'-1} a)_q^{n'}} \\
&= \frac{1 - q^{n'}}{q - 1} \frac{q^{-n'}}{(x - q^{-1} a)(x - q^{-n'-1} a)_q^{n'}} \\
&= \frac{q^{-n'} - 1}{q - 1} \frac{1}{(x - q^{-n'-1} a)_q^{n'+1}} \\
&= \frac{q^n - 1}{q - 1}(x-a)_q^{n-1},
\end{aligned}
$$

as desired.   □

Proposition 3.3 cannot be directly applied to find the q-derivatives of

$$\frac{1}{(x-a)_q^n}, \quad (a-x)_q^n, \quad \frac{1}{(a-x)_q^n},$$

because, for example, $(a-x)^n_q \neq (-1)^n(x-a)^n_q$. Instead, for $n \geq 1$,

$$
\begin{aligned}
(a-x)^n_q &= (a-x)(a-qx)(a-q^2a)\cdots(a-q^{n-1}x) \\
&= (a-x)\cdot q(q^{-1}a-x)\cdot q^2(q^{-2}-x)\cdots q^{n-1}(q^{-n+1}a-x) \\
&= (-1)^n q^{n(n-1)/2}(x-q^{-n+1}a)\cdots(x-q^{-2}a)(x-q^{-1}a)(x-a),
\end{aligned}
$$

or

$$
(a-x)^n_q = (-1)^n q^{n(n-1)/2}(x-q^{-n+1}a)^n_q. \tag{3.9}
$$

Obviously, (3.9) is true for $n=0$, and it is straightforward to verify that it is true for $n<0$.

Let us end this chapter by finding the q-derivatives of the three functions above. By (3.7), we have

$$
D_q \frac{1}{(x-a)^n_q} = D_q \frac{1}{(x-q^{-n}(q^n a))^n_q} = D_q(x-q^n a)^{-n}_q.
$$

Using (3.9) twice, we have

$$
\begin{aligned}
D_q(a-x)^n_q &= (-1)^n q^{n(n-1)/2}\cdot [n](x-q^{-n+1}a)^{n-1}_q \\
&= -[n]q^{n-1}\cdot(-1)^{n-1}q^{(n-1)(n-2)/2}\left(x-q^{-n+2}(q^{-1}a)\right)^{n-1}_q \\
&= -[n]q^{n-1}(q^{-1}a-x)^{n-1}_q = -[n](a-qx)^{n-1}_q.
\end{aligned}
$$

Finally, we use the quotient rule (1.13) and get

$$
D_q \frac{1}{(a-x)^n_q} = -\frac{-[n](a-qx)^{n-1}_q}{(a-x)^n_q(a-qx)^n_q} = \frac{[n]}{(a-x)^n_q(a-q^{n+1}x)}.
$$

To conclude, for any integer n, we have

$$
D_q \frac{1}{(x-a)^n_q} = [-n](x-q^n a)^{-n-1}_q, \tag{3.10}
$$

$$
D_q(a-x)^n_q = -[n](a-qx)^{n-1}_q, \tag{3.11}
$$

$$
D_q \frac{1}{(a-x)^n_q} = \frac{[n]}{(a-x)^{n+1}_q}. \tag{3.12}
$$

4

q-Taylor's Formula for Polynomials

As has been shown in the previous chapter, $P_n(x) = (x-a)_q^n/[n]!$ satisfies the three requirements of Theorem 2.1 with respect to the linear operator D_q. Therefore, we now obtain the q-version of Taylor's formula.

Theorem 4.1. *For any polynomial $f(x)$ of degree N and any number c, we have the following q-Taylor expansion:*

$$f(x) = \sum_{j=0}^{N} (D_q^{j} f)(c) \frac{(x-c)_q^j}{[j]!}. \tag{4.1}$$

Example. Consider $f(x) = x^n$ and $c = 1$, where n is a positive integer. For $j \le n$, we have

$$
\begin{aligned}
\left(D_q^{j} f\right)(x) &= [n] x^{n-1} = [n][n-1] x^{n-2} = \cdots \\
&= [n][n-1] \cdots [n-j+1] x^{n-j}, \tag{4.2}
\end{aligned}
$$

and hence,

$$\left(D_q^{j} f\right)(1) = [n][n-1] \cdots [n-j+1]. \tag{4.3}$$

The q-Taylor formula for x^n about $x = 1$ then gives

$$x^n = \sum_{j=0}^{n} \frac{[n] \cdots [n-j+1]}{[j]!} (x-1)_q^j = \sum_{j=0}^{n} \begin{bmatrix} n \\ j \end{bmatrix} (x-1)_q^j, \tag{4.4}$$

where

$$\begin{bmatrix} n \\ j \end{bmatrix} = \frac{[n][n-1] \cdots [n-j+1]}{[j]!} = \frac{[n]!}{[j]![n-j]!} \tag{4.5}$$

are called *q-binomial coefficients*. We will give a nice combinatorial interpretation of equation (4.4) in Chapter 7.

Note that as $q \to 1$, the q-binomial coefficients reduce to the ordinary binomial coefficients and (4.4) becomes a result of the ordinary binomial formula. The properties of the q-binomial coefficients will be examined in the next two chapters.

5

Gauss's Binomial Formula and a Noncommutative Binomial Formula

In this chapter we will encounter two binomial formulas involving q-binomial coefficients. Let us first consider an example similar to the one given in the previous chapter.

Example. Let n be a nonnegative integer and a be a number. Let us expand $f(x) = (x + a)_q^n$ about $x = 0$ using q-Taylor's formula. As in (4.2), for $j \leq n$ we have

$$\left(D_q{}^j f\right)(x) = [n][n - 1] \cdots [n - j + 1](x + a)_q^{n-j}. \tag{5.1}$$

Recall that

$$(x + a)_q^m = (x + a)(x + qa) \cdots (x + q^{m-1}a),$$

so, with $x = 0$, the right-hand side gives $(a)(qa) \cdots (q^{m-1}a) = q^{m(m-1)/2}a^m$. Apply this to (5.1) to get for $j \leq n$,

$$\left(D_q{}^j f\right)(0) = [n][n - 1] \cdots [n - j + 1]q^{(n-j)(n-j-1)/2}a^{n-j}. \tag{5.2}$$

Thus, the q-Taylor formula gives

$$(x + a)_q^n = \sum_{j=0}^{n} \begin{bmatrix} n \\ j \end{bmatrix} q^{(n-j)(n-j-1)/2}a^{n-j}x^j. \tag{5.3}$$

We can improve the expression a little bit if we replace j by $n - j$. From the definition of q-binomial coefficients (4.5), we have, similar to the usual binomial coefficients,

$$\begin{bmatrix} n \\ n - j \end{bmatrix} = \frac{[n]!}{[j]![n - j]!} = \begin{bmatrix} n \\ j \end{bmatrix}. \tag{5.4}$$

Therefore, (5.3) is equivalent to

$$(x + a)_q^n = \sum_{j=0}^{n} \begin{bmatrix} n \\ j \end{bmatrix} q^{j(j-1)/2} a^j x^{n-j}. \tag{5.5}$$

Formula (5.5) is called *Gauss's binomial formula*. It will be useful in the subsequent chapters.

Now we turn to another (though related) topic. As we all know, the multiplication of real numbers is commutative, i.e., $xy = yx$. However, when more general multiplication is concerned, such as matrix multiplication or composition of operators, commutativity may no longer be true. Consider the following example.

Example. Let $\hat{x}$ and $\hat{M}_q$ be the linear operators on the space of polynomials whose actions on a polynomial $f(x)$ are

$$\hat{x}[f(x)] = xf(x), \quad \hat{M}_q[f(x)] = f(qx). \tag{5.6}$$

Then for any $f(x)$ we have

$$\hat{M}_q\hat{x}[f(x)] = \hat{M}_q[xf(x)] = qxf(qx) = q\hat{x}\hat{M}_q[f(x)],$$

so

$$\hat{M}_q\hat{x} = q\hat{x}\hat{M}_q. \tag{5.7}$$

Theorem 5.1 below introduces a noncommutative binomial formula involving two elements satisfying a special commutation relation like (5.7).

Theorem 5.1. *If $yx = qxy$, where q is a number commuting with both x and y, then*

$$(x + y)^n = \sum_{j=0}^{n} \begin{bmatrix} n \\ j \end{bmatrix} x^j y^{n-j}. \tag{5.8}$$

Proof. Our proof is by induction on n. Equation (5.8) is obviously true for $n = 1$. Noting that $y^k x = qy^{k-1}xy = q^2 y^{k-2}xy^2 = \ldots = q^k xy^k$, we compute

$$
\begin{aligned}
(x+y)^{n+1} \;&=\; (x+y)^n(x+y) = \left(\sum_{j=0}^{n}\begin{bmatrix} n \\ j \end{bmatrix} x^j y^{n-j}\right)(x+y) \\[2mm]
&=\; \sum_{j=0}^{n}\begin{bmatrix} n \\ j \end{bmatrix} x^j y^{n-j} x + \sum_{j=0}^{n}\begin{bmatrix} n \\ j \end{bmatrix} x^j y^{n-j+1} \\[2mm]
&=\; \sum_{j=0}^{n}\begin{bmatrix} n \\ j \end{bmatrix} x^j (q^{n-j} x y^{n-j}) + \sum_{j=0}^{n}\begin{bmatrix} n \\ j \end{bmatrix} x^j y^{n-j+1} \\[2mm]
&=\; \sum_{j=1}^{n+1} q^{n-j+1}\begin{bmatrix} n \\ j-1 \end{bmatrix} x^j y^{n-j+1} + \sum_{j=0}^{n}\begin{bmatrix} n \\ j \end{bmatrix} x^j y^{n-j+1} \\[2mm]
&=\; y^{n+1} + \sum_{j=1}^{n}\left(q^{n-j+1}\begin{bmatrix} n \\ j-1 \end{bmatrix} + \begin{bmatrix} n \\ j \end{bmatrix}\right) x^j y^{n-j+1} + x^{n+1} \\[2mm]
&=\; \sum_{j=0}^{n+1}\begin{bmatrix} n+1 \\ j \end{bmatrix} x^j y^{n+1-j},
\end{aligned}
$$

where we have used the q-Pascal rule (6.3), to be discussed in the next chapter. The theorem has thus been proved. $\square$

6

Properties of q-Binomial Coefficients

Let us examine some properties of the q-binomial coefficients, defined by (4.5), with n and j being nonnegative integers and $n \geq j$. Because we will recover the ordinary binomial coefficients if we take $q \to 1$, we expect their q-analogues to have similar properties. Firstly, as already remarked in (5.4),

$$\begin{bmatrix} n \\ j \end{bmatrix} = \frac{[n]!}{[j]!\,[n-j]!} = \begin{bmatrix} n \\ n-j \end{bmatrix} \tag{6.1}$$

follows exactly the classical result. However, the correspondence is more subtle for another identity of binomial coefficients, the Pascal rule:

$$\binom{n}{j} = \binom{n-1}{j-1} + \binom{n-1}{j}, \quad 1 \leq j \leq n-1.$$

For example,

$$\begin{bmatrix} 2 \\ 1 \end{bmatrix} = 1 + q \neq 2 = \begin{bmatrix} 1 \\ 0 \end{bmatrix} + \begin{bmatrix} 1 \\ 1 \end{bmatrix}.$$

Proposition 6.1. *There are two q-Pascal rules, namely,*

$$\begin{bmatrix} n \\ j \end{bmatrix} = \begin{bmatrix} n-1 \\ j-1 \end{bmatrix} + q^j \begin{bmatrix} n-1 \\ j \end{bmatrix} \tag{6.2}$$

and

$$\begin{bmatrix} n \\ j \end{bmatrix} = q^{n-j} \begin{bmatrix} n-1 \\ j-1 \end{bmatrix} + \begin{bmatrix} n-1 \\ j \end{bmatrix}, \tag{6.3}$$

where $1 \leq j \leq n-1$.

Proof. Because for any $1 \leq j \leq n-1$,

$$
\begin{aligned}
[n] &= 1 + q + \cdots + q^{n-1} \\
&= (1 + q + \cdots + q^{j-1}) + q^j(1 + q + \cdots + q^{n-j-1}) \\
&= [j] + q^j[n-j],
\end{aligned}
$$

we have

$$
\begin{aligned}
\begin{bmatrix} n \\ j \end{bmatrix} &= \frac{[n]!}{[j]![n-j]!} = \frac{[n-1]![n]}{[j]![n-j]!} \\
&= \frac{[n-1]!([j] + q^j[n-j])}{[j]![n-j]!} \\
&= \frac{[n-1]!}{[j-1]![n-j]!} + q^j \frac{[n-1]!}{[j]![n-j-1]!} \\
&= \begin{bmatrix} n-1 \\ j-1 \end{bmatrix} + q^j \begin{bmatrix} n-1 \\ j \end{bmatrix},
\end{aligned}
$$

which is (6.2). The symmetric property of the coefficients (6.1) gives us the other identity, because

$$
\begin{aligned}
\begin{bmatrix} n \\ j \end{bmatrix} &= \begin{bmatrix} n \\ n-j \end{bmatrix} = \begin{bmatrix} n-1 \\ n-j-1 \end{bmatrix} + q^{n-j} \begin{bmatrix} n-1 \\ n-j \end{bmatrix} \\
&= \begin{bmatrix} n-1 \\ j \end{bmatrix} + q^{n-j} \begin{bmatrix} n-1 \\ j-1 \end{bmatrix}. \quad \square
\end{aligned}
$$

Corollary 6.1. *Each q-binomial coefficient is a polynomial in q of degree $j(n-j)$, with 1 as the leading coefficient.*

Proof. For any nonnegative integer n,

$$
\begin{bmatrix} n \\ 0 \end{bmatrix} = \begin{bmatrix} n \\ n \end{bmatrix} = 1,
$$

which is of course a polynomial. Using Proposition 6.1 and induction on n, for any $1 \leq j \leq n-1$, $\begin{bmatrix} n \\ j \end{bmatrix}$ is the sum of two polynomials, thus is itself a polynomial.

By definitions (4.5) and (1.9), the explicit expression of a q-binomial coefficient is

$$
\begin{bmatrix} n \\ j \end{bmatrix} = \frac{(q^n - 1)(q^{n-1} - 1) \cdots (q^{n-j+1} - 1)}{(q^j - 1)(q^{j-1} - 1) \cdots (q - 1)}. \tag{6.4}
$$

Since both the numerator and denominator of (6.4) are polynomials in q with leading coefficient 1, so is their quotient. Finally, the degree of $\begin{bmatrix} n \\ j \end{bmatrix}$ in q is the difference of the degrees of the numerator and denominator, which is $[n + (n-1) + \cdots + (n-j+1)] - [j + (j-1) + \cdots + 1] = (n-j) + (n-j) + \cdots + (n-j) = j(n-j)$. $\quad \square$

Another fact can be deduced from the explicit expression (6.4) of the q-binomial coefficient. Knowing that it is a polynomial in q of degree $j(n-j)$, we let

$$a_0 + a_1 q + \cdots + a_{j(n-j)-1} q^{j(n-j)-1} + a_{j(n-j)} q^{j(n-j)}$$
$$= \frac{(q^n - 1)(q^{n-1} - 1) \cdots (q^{n-j+1} - 1)}{(q^j - 1)(q^{j-1} - 1) \cdots (q - 1)}.$$

If we replace q by $1/q$ and multiply both sides by $q^{j(n-j)}$, it is easy to check that the right-hand side will be unchanged, while the left-hand side,

$$a_0 q^{j(n-j)} + a_1 q^{j(n-j)-1} + \cdots + a_{j(n-j)-1} q + a_{j(n-j)},$$

has the sequence of coefficients a_i reversed in order. By comparing coefficients, we observe that the coefficients in the polynomial expression of $\begin{bmatrix} n \\ j \end{bmatrix}$ are symmetric, i.e., $a_i = a_{j(n-j)-i}$.

Like the ordinary binomial coefficients, the q-binomial coefficients also have combinatorial interpretations. Here is one of them, and another one will be given in the next chapter.

Theorem 6.1. *Let $A_n = \{1, 2, \ldots, n\}$ and let $A_{n,j}$ be the collection of all subsets of A_n with j elements, $0 \le j \le n$. Then*

$$\begin{bmatrix} n \\ j \end{bmatrix} = \sum_{S \in A_{n,j}} q^{w(S)-j(j+1)/2}, \quad \text{where } w(S) = \sum_{s \in S} s. \tag{6.5}$$

Proof. We will prove the theorem by induction on n. First, consider $n = 1$, $j = 0, 1$. For $j = 0$, $A_{1,0} = \{\phi\}$ and $w(\phi) = 0$. Thus, the right-hand side of (6.5) equals unity, agreeing with the left-hand side. For $j = 1$, the only element of $A_{1,1}$ is $A_1 = \{1\}$, and $w(\{1\}) = 1$. Again, the right-hand side equals unity and agrees with the left-hand side.

Assume that (6.5) holds for $1 \le n \le m - 1$, where $m \ge 2$, and consider $n = m$. The case $j = 0$ is similar to that for $n = 1$ described above. For $j \ge 1$, write $A_{m,j} = B \cup B'$, where $B = \{S \in A_{m,j} | m \notin S\}$ and $B' = \{S \in A_{m,j} | m \in S\}$. The sets in B are all the j-element subsets of A_{m-1}, i.e., $B = A_{m-1,j}$. The sets in B' each with the element "m" removed are all the $(j-1)$-element subsets of A_{m-1}. Hence, the right-hand side of

(6.5) becomes

$$\sum_{S \in \mathcal{B}} q^{w(S)-j(j+1)/2} + \sum_{S \in \mathcal{B}'} q^{w(S)-j(j+1)/2}$$

$$= \sum_{S \in \mathcal{A}_{m-1,j}} q^{w(S)-j(j+1)/2}$$

$$+ \sum_{S \in \mathcal{A}_{m-1,j-1}} q^{\left(w(S)+m\right)-j(j+1)/2}$$

$$= \sum_{S \in \mathcal{A}_{m-1,j}} q^{w(S)-j(j+1)/2}$$

$$+ \sum_{S \in \mathcal{A}_{m-1,j-1}} q^{w(S)-j(j-1)/2} \cdot q^{m-j}$$

$$= \begin{bmatrix} m-1 \\ j \end{bmatrix} + q^{m-j} \begin{bmatrix} m-1 \\ j-1 \end{bmatrix} = \begin{bmatrix} m \\ j \end{bmatrix}.$$

The last line follows from one of the q-Pascal rules (6.3). By induction on j, (6.5) is true for $0 \leq j \leq m$. Finally, induction on n completes the proof. $\square$

For future use, note that the definition of the q-binomial coefficient can be generalized in a way similar to its ordinary counterpart, using (3.8):

$$\begin{bmatrix} \alpha \\ j \end{bmatrix} = \frac{[\alpha][\alpha-1]\cdots[\alpha-j+1]}{[j]!}, \tag{6.6}$$

where α is any number and j is a nonnegative integer.

7

q-Binomial Coefficients and Linear Algebra over Finite Fields

In this chapter we explain an important combinatorial meaning of the q-binomial coefficients.

Theorem 7.1. *If q is the order of a finite field $\mathbb{F}_q$ (hence, q is a prime power), then*

$$\begin{bmatrix} n \\ j \end{bmatrix} = \begin{array}{l} \text{number of } j\text{-dimensional subspaces in the} \\ n\text{-dimensional vector space } \mathbb{F}_q^n. \end{array} \qquad (7.1)$$

Before proving the theorem, let us very briefly review some basic concepts of linear algebra. A collection of vectors in a vector space V over a field F is a subspace if it contains the zero vector and it is closed under vector addition and scalar multiplication. The dimension of a subspace, if finite, is given by the number of vectors in a basis for the subspace, which is a collection of linearly independent vectors that span the whole subspace. The only zero-dimensional subspace is $\{0\}$. A one-dimensional subspace is spanned by one nonzero vector, $\{av | v \neq 0, a \in F\}$, a two-dimensional subspace is spanned by two linearly independent vectors, $\{av_1 + bv_2 | v_1, v_2 \text{ linearly independent}, a, b \in F\}$, and so on. The vector space $\mathbb{F}_q^n$ consists of all n-tuples, or n-component vectors,

$$(a_1, a_2, \ldots, a_n),$$

where each a_i is an element of the finite field $\mathbb{F}_q$. Since $|\mathbb{F}_q| = q$, there are q^n such n-tuples, or $|\mathbb{F}_q^n| = q^n$.

Proof of Theorem 7.1. Let $V = \mathbb{F}_q^n$. For $j = 0$, $\begin{bmatrix} n \\ 0 \end{bmatrix} = 1$ and there is only one zero-dimensional subspace of V, so this case is proved.

For $j \geq 1$, to obtain a j-dimensional subspace, we choose j linearly independent vectors in V to form a basis. The first one, v_1, can be any one of the $q^n - 1$ nonzero vectors. The second one, v_2, can be any vector not in the subspace spanned by v_1. Since a one-dimensional subspace of V has q elements, there are $q^n - q$ different choices for the second basis vector. Then, the number of choices for the third one, v_3, is $q^n - q^2$, because it can be any vector not in the two-dimensional subspace spanned by v_1 and v_2, which has q^2 elements. In general, after the ith basis vector is picked, the number of vectors in the subspace spanned by the first i basis vectors is q^i, and we are left with $q^n - q^i$ choices for the $(i+1)$th one. Thus, we have

$$(q^n - 1)(q^n - q)(q^n - q^2) \cdots (q^n - q^{j-1}) \tag{7.2}$$

different ways to choose j linearly independent vectors in $\mathbb{F}_q^n$.

However, many of these j-tuples span the same subspace. We have to divide the expression in (7.2) by the number of different possible choices of basis of a particular j-dimensional subspace. But it is essentially the same number in (7.2), with n replaced by j. Therefore, the number of *different* j-dimensional subspaces is

$$
\begin{aligned}
&\frac{(q^n - 1)(q^n - q)(q^n - q^2) \cdots (q^n - q^{j-1})}{(q^j - 1)(q^j - q)(q^j - q^2) \cdots (q^j - q^{j-1})} \\
&= \frac{q \cdot q^2 \cdots q^{j-1} \cdot (q^n - 1)(q^{n-1} - 1) \cdots (q^{n-j+1} - 1)}{q \cdot q^2 \cdots q^{j-1} \cdot (q^j - 1)(q^{j-1} - 1) \cdots (q - 1)} \\
&= \begin{bmatrix} n \\ j \end{bmatrix},
\end{aligned}
$$

according to (6.4).   □

Like the Pascal rule, many identities involving binomial coefficients have their q-analogues. Imagine that we have $m + n$ balls, and they are placed into two groups, one with m and one with n of them. Each way of choosing k balls from all $m + n$ of them corresponds in a one-to-one manner to a way of choosing j balls from the group with m balls and choosing $k - j$ balls from the group of n balls, with j running from 0 to k. Hence, we have the following identity of binomial coefficients:

$$\binom{m+n}{k} = \sum_{j=0}^{k} \binom{m}{j}\binom{n}{k-j}. \tag{7.3}$$

Example. Obtain a q-analogue of the identity (7.3) using the combinatorial interpretation of q-binomial coefficients as stated in Theorem 7.1.

Let $V = \mathbb{F}_q^{m+n}$ and let $V_m \subset V$ be a fixed subspace with $\dim V_m = m$. We would like to obtain an identity by counting the number of k-dimensional subspaces in V in two ways. First, by Theorem 7.1, we know that this number is $\begin{bmatrix} m+n \\ k \end{bmatrix}$.

On the other hand, let W be a k-dimensional subspace of V. As the intersection of two subspaces, $W \cap V_m$ is also a subspace, of dimension j, which is between 0 and k. We may then regard each W as being extended from a j-dimensional subspace of V_m. Suppose such a subspace $W' \subset V_m$, $\dim W' = j$, has been chosen. We now append $k - j$ linearly independent vectors $(v_1, v_2, \ldots, v_{k-j})$ to W' to form W: v_1 can be chosen from the $q^{m+n} - q^m$ vectors not in V_m, v_2 can be chosen from the $q^{m+n} - q^{m+1}$ vectors not in the subspace spanned by V_m and v_1, etc. By the same argument as in the proof of Theorem 7.1, there are

$$\left(q^{m+n} - q^m\right)\left(q^{m+n} - q^{m+1}\right) \cdots \left(q^{m+n} - q^{m+k-j-1}\right) \tag{7.4}$$

different ways to append $k - j$ linearly independent vectors to W'.

Again, we have to count the number of different ways of extending W' to a single W. Since $\dim W = k$ and $\dim W' = j$, the number is

$$(q^k - q^j)(q^k - q^{j+1}) \cdots (q^k - q^{k-1}), \tag{7.5}$$

according to similar arguments. Therefore, the number of *different* W obtained by extending from a given W' is

$$
\begin{aligned}
&\frac{(q^{m+n} - q^m)(q^{m+n} - q^{m+1}) \cdots (q^{m+n} - q^{m+k-j-1})}{(q^k - q^j)(q^k - q^{j+1}) \cdots (q^k - q^{k-1})} \\
=\ &\frac{q^m \cdot q^{m+1} \cdots q^{m+k-j-1} \cdot (q^n - 1)(q^{n-1} - 1) \cdots (q^{n-k+j+1} - 1)}{q^j \cdot q^{j+1} \cdots q^{k-1} \cdot (q^{k-j} - 1)(q^{k-j-1} - 1) \cdots (q - 1)} \\
=\ &q^{(k-j)(m-j)} \begin{bmatrix} n \\ k-j \end{bmatrix}.
\end{aligned}
$$

Because there are $\begin{bmatrix} m \\ j \end{bmatrix}$ different choices for W' and any two of them generate distinct W, we obtain the identity

$$\begin{bmatrix} m+n \\ k \end{bmatrix} = \sum_{j=0}^{k} q^{(k-j)(m-j)} \begin{bmatrix} m \\ j \end{bmatrix} \begin{bmatrix} n \\ k-j \end{bmatrix}, \tag{7.6}$$

which is a q-analogue of (7.3).

Example. Recall from (4.4) the q-Taylor expansion of $f(x) = x^n$ about $x = 1$:

$$x^n = \sum_{j=0}^{n} \begin{bmatrix} n \\ j \end{bmatrix} (x - 1)_q^j.$$

As promised earlier, we will now prove this expansion again using combinatorial arguments. Our strategy is to show that the identity holds if $x = q^m$, where m is any positive integer. Since both sides of the identity are polynomials, equality at infinitely many points ensures equality at all points. (If f, g are polynomials and $f(x) = g(x)$ for infinitely many values of x, the polynomial $h(x) = f(x) - g(x)$ has infinitely many zeros, which is possible only if h is identically zero.)

Let n, m be positive integers and S be the set of all linear transformations from $A = \mathbb{F}_q^n$ to $B = \mathbb{F}_q^m$. Suppose $\{e_1, \ldots, e_n\}$ is a basis of A. Since, given any $T \in S$, $T(e_k)$ can be any of the q^m vectors in B, for each $1 \le k \le n$, and together they uniquely determine T, the number of elements $|S|$ of S is thus $(q^m)^n$, which is the left-hand side of (4.4) when $x = q^m$. On the other hand, we can write

$$|S| = \sum_{j=0}^{n} (\text{number of elements in } S \text{ of rank } j).$$

We desire to show that the jth summand is $\begin{bmatrix} n \\ j \end{bmatrix} (q^m - 1)_q^j$. Note that the rank of T cannot be larger than m, which agrees with the fact that $(q^m - 1)_q^j = 0$ when $j > m$. Thus, we consider only $j \le m$.

Here we use some facts about linear transformations. That T has rank j means that $W = T(A) \subset B$ is a j-dimensional subspace, and A can be decomposed as a direct sum of two subspaces, $A = V \oplus K$, where $\dim V = j$ and $\dim K = n - j$, such that T maps V onto W in a one-to-one fashion and $K = \{v \in A \,|\, T(v) = 0\}$. In other words, any vector in A may be represented as a sum of two vectors in a unique way, so that one is in V and the other in K. (To see why such a decomposition is possible, choose $u_1, \ldots, u_j$ in A such that their images form a basis for W. The linear independence of their images implies their own linear independence. Let V be the space spanned by u_i. For any $v \in A$, $T(v) \in W$. Since $T(u_i)$ is a basis of W, $T(v) = \sum a_i T(u_i)$ for some a_i. Let $v' = \sum a_i u_i$. Then, $v' \in V$ and $T(v - v') = 0$; thus $v - v' \in K$. That K is a subspace is easy to show. Since $T(V) = W$ and both V and W contain q^j vectors, T is one-to-one on V, and thus v' is the only vector in V such that $T(v) = T(v')$, implying the uniqueness of the decomposition.)

From another perspective, we may specify T by choosing the subspaces $V \subset A$ and $W \subset B$ and the way V is mapped into W. By Theorem 7.1, the number of choices for V and W is $\begin{bmatrix} n \\ j \end{bmatrix} \begin{bmatrix} m \\ j \end{bmatrix}$. Now, suppose V and W are given, and let $\{u_1, \ldots, u_j\}$ be a basis of V. Keeping in mind that T is one-to-one, we know that $T(u_1)$ can be any of the $q^j - 1$ nonzero vectors in W, $T(u_2)$ can be any of the $q^j - q$ vectors in W not in the span of $T(u_1)$, $T(u_3)$ can be any of the $q^j - q^2$ vectors in W not in the span of $T(u_1)$ and $T(u_2)$, and so on. Hence, there are $(q^j - 1)(q^j - q) \cdots (q^j - q^{j-1})$ ways to map V into W bijectively. Therefore, the number of elements in S of rank

j is

$$\begin{bmatrix} n \\ j \end{bmatrix} \begin{bmatrix} m \\ j \end{bmatrix} \prod_{i=0}^{j-1} (q^j - q^i) = \begin{bmatrix} n \\ j \end{bmatrix} \prod_{i=0}^{j-1} \left(\frac{q^{m-i} - 1}{q^{j-i} - 1} \right) \prod_{i=0}^{j-1} (q^j - q^i)$$

$$= \begin{bmatrix} n \\ j \end{bmatrix} \prod_{i=0}^{j-1} (q^m - q^i) = \begin{bmatrix} n \\ j \end{bmatrix} (q^m - 1)_q^j,$$

as desired.

Theorem 7.1 also tells us that the total number of subspaces of the vector space $\mathbb{F}_q^n$ is given by

$$G_n = \sum_{j=0}^{n} \begin{bmatrix} n \\ j \end{bmatrix}, \tag{7.7}$$

which is called the nth *Galois number*. If the q-binomial coefficients are replaced by ordinary ones, the sum is exactly 2^n. However, the q-calculus case is not as easy to calculate explicitly. Instead, the Galois numbers may be computed recursively, as was shown by Goldman and Rota.

Proposition 7.1. *The Galois numbers satisfy the following recursive relation:*

$$G_{n+1} = 2G_n + (q^n - 1)G_{n-1}, \tag{7.8}$$

with $G_0 = 1$ and $G_1 = 2$.

Proof. Let $P_n(x) = (x - 1)_q^n$. The trick we are going to use is to define a linear function L on the space of polynomials such that

$$L\{P_n(x)\} = 1 \tag{7.9}$$

for any nonnegative integer n. Such a linear function exists because the polynomials $(x - a)_q^n$ are linearly independent (for different n). If we apply L to both sides of (4.4), we have

$$L\{x^n\} = \sum_{j=0}^{n} \begin{bmatrix} n \\ j \end{bmatrix} L\{P_j(x)\} = \sum_{j=0}^{n} \begin{bmatrix} n \\ j \end{bmatrix} = G_n. \tag{7.10}$$

To exploit the linear property of L, note that $P_{n+1}(x) = (x - q^n)P_n(x) = xP_n(x) - q^n P_n(x)$, hence

$$L\{xP_n(x)\} = L\{P_{n+1}(x)\} + q^n L\{P_n(x)\} = 1 + q^n. \tag{7.11}$$

On the other hand, from $D_q P_n(x) = [n]P_{n-1}(x)$, we have

$$1 + q^n = 2L\{P_n(x)\} + (q - 1)L\{D_q P_n(x)\}. \tag{7.12}$$

Equating (7.11) and (7.12), we obtain

$$L\{xP_n(x)\} = 2L\{P_n(x)\} + (q - 1)L\{D_q P_n(x)\}, \tag{7.13}$$

which is true for any $n \geq 0$. Since any polynomial can be expressed as a linear combination of $P_n(x)$, we can replace $P_n(x)$ in (7.13) by any polynomial. In particular, if we replace it by x^n, we get

$$
\begin{aligned}
L\{x^{n+1}\} &= 2L\{x^n\} + (q-1)L\{[n]x^{n-1}\} \\
&= 2L\{x^n\} + (q^n - 1)L\{x^{n-1}\}.
\end{aligned}
$$

According to (7.10), this proves (7.8). $\square$

Furthermore, if we choose another linear function L' such that $L'\{P_n\} = t^n$, we may obtain the following recursive formula in a similar manner:

$$
L'\{x^{n+1}\} = (t+1)L'\{x^n\} + t(q^n - 1)L'\{x^{n-1}\}.
$$

And if we define the sequence

$$
f_n(t) = \sum_{j=0}^{n} \left[\begin{array}{c} n \\ j \end{array} \right] t^j
$$

of polynomials in t, we have $f_n(t) = L'\{x^n\}$, and thus

$$
f_{n+1}(t) = (t+1)f_n(t) + (q^n - 1)t f_{n-1}(t), \qquad n \geq 1.
$$

Note that $G_n = f_n(1)$ and putting $t = 1$ above recovers Proposition 7.1. When $t = -1$, the recursive relation is particularly simple:

$$
f_{n+1}(-1) = (1 - q^n)f_{n-1}(-1), \qquad n \geq 1.
$$

Since $f_0(-1) = 1$ and $f_1(-1) = 0$, we have

$$
\sum_{j=0}^{2m} (-1)^j \left[\begin{array}{c} 2m \\ j \end{array} \right] = (1 - q^{2m-1})(1 - q^{2m-3}) \cdots (1 - q), \tag{7.14}
$$

and

$$
\sum_{j=0}^{2m+1} (-1)^j \left[\begin{array}{c} 2m+1 \\ j \end{array} \right] = 0, \tag{7.15}
$$

for any $m \geq 0$. These two identities were first discovered by Gauss. The present proof is due to Goldman and Rota.

8
q-Taylor's Formula for Formal Power Series and Heine's Binomial Formula

We now begin to apply what we have learned so far, particularly q-Taylor's formula (4.1), to study identities involving infinite sums and products. In order to do this, we first have to remark that the generalized Taylor formula (2.2) about $a = 0$, and hence the q-Taylor formula (4.1) about $c = 0$, apply not only to polynomials, but also to formal power series. A formal power series, of the form

$$f(x) = \sum_{k=0}^{\infty} c_k x^k,$$

may be thought of as a polynomial of infinite degree. It is "formal" because often we do not worry about whether the series converges or not, and we can operate on (for example, differentiate) the series formally. We have to assume a and c to be zero in order to avoid divergence problems. Of course, $f(0) = c_0$ by definition.

The q-derivative of the formal power series $f(x)$ is, of course, $D_q f(x) = \sum_{k=0}^{\infty} [k] c_k x^{k-1}$. Hence we have

$$[k]! c_k = (D_q^k f(x))(0).$$

It follows, in particular, that if two formal power series converge in some neighborhood of 0 to the same function, then they are equal.

Theorem 8.1. *Suppose D is a linear operator on the space of formal power series and $\{P_0(x), P_1(x), P_2(x), \ldots\}$ is a sequence of polynomials such that the three conditions in Theorem 2.1 are satisfied for $a = 0$. Then, any* *formal power series $f(x)$ can be expressed as a generalized Taylor's series (2.2) about $x = 0$.*

Corollary 8.1. *Any formal power series $f(x)$ can be expressed as a q-Taylor series (4.1) about $x = 0$.*

Proof of Theorem 8.1. It is easy to see by induction on n that in the case $a = 0$ the three conditions of Theorem 2.1 imply that $P_n(x) = a_k x^k$, where the a_k are nonzero numbers. Hence for any formal power series $f(x)$, we have

$$f(x) = \sum_{j=0}^{\infty} c_j P_j(x)$$

for some constants c_j. Applying D k times and putting $x = 0$ yields $c_k = (D^k f)(0)$, which completes the proof. □

Example. Consider the function $f(x) = 1/(1-x)_q^n$. Using long division, we can see that $f(x)$ is a formal power series. Let us expand $f(x)$ using q-Taylor's formula about $x = 0$. From (3.12), we have

$$D_q f(x) = D_q \frac{1}{(1-x)_q^n} = \frac{[n]}{(1-x)_q^{n+1}},$$

and, by induction,

$$D_q^j f(x) = \frac{[n][n+1]\cdots[n+j-1]}{(1-x)_q^{n+j}}.$$

Hence, $(D_q^j f)(0) = [n][n+1]\cdots[n+j-1]$ for any $j \geq 1$, and therefore,

$$\frac{1}{(1-x)_q^n} = 1 + \sum_{j=1}^{\infty} \frac{[n][n+1]\cdots[n+j-1]}{[j]!} x^j, \tag{8.1}$$

which is the q-analogue of Taylor's expansion of $f(x) = 1/(1-x)^n$ in ordinary calculus. Formula (8.1) is called *Heine's binomial formula*.

9
Two Euler's Identities and Two q-Exponential Functions

Now we have two binomial formulas, namely Gauss's binomial formula (5.5) (with x and a replaced by 1 and x respectively)

$$(1+x)_q^n = \sum_{j=0}^{n} q^{j(j-1)/2} \begin{bmatrix} n \\ j \end{bmatrix} x^j,$$

and Heine's binomial formula (8.1)

$$\frac{1}{(1-x)_q^n} = \sum_{j=0}^{\infty} \frac{[n][n+1]\cdots[n+j-1]}{[j]!} x^j.$$

What if we let $n \to \infty$ in both formulas? In the ordinary calculus, i.e., $q = 1$, the answer is not very interesting. It is either infinitely large or infinitely small, depending on the value of x. However, it is different in quantum calculus, because, for example, when $|q| < 1$, the infinite product $(1+x)_q^\infty = (1+x)(1+qx)(1+q^2x)\cdots$ converges to some finite limit. Moreover, if we assume $|q| < 1$, we have

$$\lim_{n\to\infty} [n] = \lim_{n\to\infty} \frac{1-q^n}{1-q} = \frac{1}{1-q} \tag{9.1}$$

and

$$\lim_{n\to\infty} \begin{bmatrix} n \\ j \end{bmatrix} = \lim_{n\to\infty} \frac{(1-q^n)(1-q^{n-1})\cdots(1-q^{n-j+1})}{(1-q)(1-q^2)\cdots(1-q^j)}.$$

Thus

$$\lim_{n \to \infty} \begin{bmatrix} n \\ j \end{bmatrix} = \frac{1}{(1-q)(1-q^2) \cdots (1-q^j)}. \tag{9.2}$$

So, the q-analogues of integers and binomial coefficients behave in a very different way when n is large as compared to their ordinary counterparts.

If we apply (9.1) and (9.2) to Gauss's and Heine's binomial formulas, we obtain, as $n \to \infty$, the following two identities of formal power series in x (assuming that $|q| < 1$):

$$(1 + x)_q^\infty = \sum_{j=0}^{\infty} q^{j(j-1)/2} \frac{x^j}{(1-q)(1-q^2) \cdots (1-q^j)}, \tag{9.3}$$

$$\frac{1}{(1-x)_q^\infty} = \sum_{j=0}^{\infty} \frac{x^j}{(1-q)(1-q^2) \cdots (1-q^j)}. \tag{9.4}$$

The two identities above relate infinite products to infinite sums. They have no classical analogues because each term in the sums has no meaning when $q = 1$. Interestingly, the two identities were discovered by Euler, who lived before Gauss and Heine. We shall call (9.3) and (9.4) Euler's first and second identities, respectively, or E1 and E2.

Let us study E2 more closely. Consider

$$\sum_{j=0}^{\infty} \frac{x^j}{(1-q)(1-q^2) \cdots (1-q^j)} = \sum_{j=0}^{\infty} \frac{\left(\frac{x}{1-q}\right)^j}{1\left(\frac{1-q^2}{1-q}\right) \cdots \left(\frac{1-q^j}{1-q}\right)}$$

$$= \sum_{j=0}^{\infty} \frac{\left(\frac{x}{1-q}\right)^j}{[j]!}, \tag{9.5}$$

which resembles Taylor's expansion of the classical exponential function:

$$e^x = \sum_{j=0}^{\infty} \frac{x^j}{j!}. \tag{9.6}$$

Definition. A q-analogue of the classical exponential function e^x is

$$e_q^x = \sum_{j=0}^{\infty} \frac{x^j}{[j]!}. \tag{9.7}$$

Then, from (9.4) and (9.5), we immediately have

$$e_q^{x/(1-q)} = \frac{1}{(1-x)_q^\infty}, \tag{9.8}$$

or

$$e_q^x = \frac{1}{\left(1 - (1-q)x\right)_q^\infty}. \tag{9.9}$$

Analogously, we can define another q-exponential function using E1.

Definition. Another q-analogue of the classical exponential function is

$$E_q^x = \sum_{j=0}^{\infty} q^{j(j-1)/2} \frac{x^j}{[j]!} = \left(1 + (1-q)x\right)_q^{\infty}. \tag{9.10}$$

Let us study some properties of the two q-exponential functions. The classical exponential function is unchanged under differentiation. Its two q-analogues have similar behavior. Since

$$D_q e_q^x = \sum_{j=0}^{\infty} \frac{D_q x^j}{[j]!} = \sum_{j=1}^{\infty} \frac{[j] x^{j-1}}{[j]!} = \sum_{j=1}^{\infty} \frac{x^{j-1}}{[j-1]!} = \sum_{j=0}^{\infty} \frac{x^j}{[j]!},$$

and,

$$\begin{aligned}
D_q E_q^x &= \sum_{j=0}^{\infty} q^{j(j-1)/2} \frac{D_q x^j}{[j]!} = \sum_{j=1}^{\infty} q^{j(j-1)/2} \frac{[j] x^{j-1}}{[j]!} \\
&= \sum_{j=1}^{\infty} q^{(j-1)(j-2)/2} q^{j-1} \frac{x^{j-1}}{[j-1]!} = \sum_{j=0}^{\infty} q^{j(j-1)/2} \frac{q^j x^j}{[j]!},
\end{aligned}$$

we have

$$D_q e_q^x = e_q^x \quad \text{and} \quad D_q E_q^x = E_q^{qx}. \tag{9.11}$$

Note that the derivative of E_q^x is not exactly itself. The results in (9.11) may also be obtained by letting $n \to \infty$ in

$$D_q \frac{1}{\left(1 - (1-q)x\right)_q^n} = \frac{(1-q)[n]}{\left(1 - (1-q)x\right)_q^{n+1}}$$

and

$$D_q \left(1 + (1-q)x\right)_q^n = (1-q)[n]\left(1 + q(1-q)x\right)_q^{n-1}.$$

How about $e_q^x e_q^y$? In general, $e_q^x e_q^y \neq e_q^{x+y}$. But the additive property of the exponentials holds if x and y satisfy the commutation relation mentioned in Chapter 5, i.e., $yx = qxy$. To see this, consider

$$\begin{aligned}
e_q^x e_q^y &= \left(\sum_{j=0}^{\infty} \frac{x^j}{[j]!}\right)\left(\sum_{k=0}^{\infty} \frac{y^k}{[k]!}\right) = \sum_{j=0}^{\infty}\sum_{k=0}^{\infty} \frac{x^j y^k}{[j]![k]!} \\
&= \sum_{j=0}^{\infty}\sum_{k=0}^{\infty} \frac{[j+k]!}{[j]![k]!} \cdot \frac{x^j y^k}{[j+k]!}.
\end{aligned}$$

If we change variables from j and k to j and $n = j+k$, then for a particular value of n, j runs from 0 to n. Using Theorem 5.1, we have

$$e_q^x e_q^y = \sum_{n=0}^{\infty}\left(\sum_{j=0}^{n} \begin{bmatrix} n \\ j \end{bmatrix} x^j y^{n-j}\right)\frac{1}{[n]!} = \sum_{n=0}^{\infty} \frac{(x+y)^n}{[n]!}.$$

Hence, we have

$$e_q^x e_q^y = e_q^{x+y} \qquad \text{if } yx = qxy. \tag{9.12}$$

Due to their commutation relation, x and y are not symmetric, and $e_q^y e_q^x \neq e_q^x e_q^y$.

Also, the two q-exponential functions are closely related. From (9.9) and (9.10), we see that

$$e_q^x E_q^{-x} = 1, \tag{9.13}$$

and, using (9.3) and (9.4) as well, we obtain

$$
\begin{aligned}
e_{1/q}^x &= \sum_{j=0}^{\infty} \frac{(1 - 1/q)^j x^j}{(1 - 1/q)(1 - 1/q^2) \cdots (1 - 1/q^j)} \\
&= \sum_{j=0}^{\infty} q^{j(j-1)/2} \frac{(1 - q)^j x^j}{(1 - q)(1 - q^2) \cdots (1 - q^j)},
\end{aligned}
$$

and thus

$$e_{1/q}^x = E_q^x. \tag{9.14}$$

10
q-Trigonometric Functions

The q-analogues of the sine and cosine functions can be defined in analogy with their well-known Euler expressions in terms of the exponential function.

Definition. The q-trigonometric functions are

$$\sin_q x = \frac{e_q^{ix} - e_q^{-ix}}{2i}, \quad \mathrm{Sin}_q x = \frac{E_q^{ix} - E_q^{-ix}}{2i}, \tag{10.1}$$

$$\cos_q x = \frac{e_q^{ix} + e_q^{-ix}}{2}. \quad \mathrm{Cos}_q x = \frac{E_q^{ix} + E_q^{-ix}}{2}. \tag{10.2}$$

From (9.14) we have $\mathrm{Sin}_q x = \sin_{1/q} x$ and $\mathrm{Cos}_q x = \cos_{1/q} x$. Also, using (9.13), we get

$$\cos_q x \mathrm{Cos}_q x = \frac{e_q^{ix} E_q^{ix} + e_q^{-ix} E_q^{-ix} + 2}{4}$$

and

$$\sin_q x \mathrm{Sin}_q x = -\frac{e_q^{ix} E_q^{ix} + e_q^{-ix} E_q^{-ix} - 2}{4}.$$

Hence, we have

$$\cos_q x \mathrm{Cos}_q x + \sin_q x \mathrm{Sin}_q x = 1, \tag{10.3}$$

which is the q-analogue of the identity $\sin^2 x + \cos^2 x = 1$. The reader is invited to try to find q-analogues of other trigonometric formulas.

To find the derivatives of the q-trigonometric functions, we apply the chain rule (1.15), where $u(x) = ix$, and use (9.11). Then, we obtain

$$D_q \sin_q x = \cos_q x, \qquad D_q \mathrm{Sin}_q x = \mathrm{Cos}_q qx, \qquad (10.4)$$

$$D_q \cos_q x = -\sin_q x, \qquad D_q \mathrm{Cos}_q x = -\mathrm{Sin}_q qx. \qquad (10.5)$$

11

Jacobi's Triple Product Identity

We recall that the two Euler identities, (9.3) and (9.4), relate infinite products and infinite sums. In this chapter, we will use them to prove an important identity first discovered by Jacobi. Several interesting applications of this identity in number theory will be explored in subsequent chapters.

Theorem 11.1. *If $|q| < 1$, we have*

$$\sum_{n=-\infty}^{\infty} q^{n^2} z^n = \prod_{n=1}^{\infty} (1 - q^{2n})(1 + q^{2n-1}z)(1 + q^{2n-1}z^{-1}), \qquad (11.1)$$

which is called Jacobi's triple product identity.

Proof (G.E. Andrews). We start with E1:

$$\prod_{n=0}^{\infty} (1 + q^n x) = \sum_{j=0}^{\infty} \frac{q^{j(j-1)/2} x^j}{(1 - q)(1 - q^2) \cdots (1 - q^j)}. \qquad (11.2)$$

If we replace q by q^2, and then x by zq, we obtain

$$\prod_{n=1}^{\infty} (1 + q^{2n-1}z) = \prod_{n=0}^{\infty} (1 + q^{2n+1}z) = \sum_{j=0}^{\infty} \frac{q^{j^2} z^j}{(1 - q^2)(1 - q^4) \cdots (1 - q^{2j})}.$$

The product in the denominator of each summand can be removed by multiplying both sides by

$$\prod_{n=1}^{\infty} (1 - q^{2n}),$$

giving

$$\prod_{n=1}^{\infty} \left(1 - q^{2n}\right)\left(1 + q^{2n-1}z\right) = \sum_{j=-\infty}^{\infty} \left(q^{j^2} z^j \prod_{n=0}^{\infty} \left(1 - q^{2n+2j+2}\right)\right). \quad (11.3)$$

Note that the summation on the right now starts from $-\infty$ instead of zero. The sum is unchanged because $(1 - q^{2n+2j+2}) = 0$ for some $n \geq 0$ if j is negative. On the other hand, we use (11.2) again, replacing the index j by k, q by q^2, and then x by $-q^{2j+2}$, to obtain

$$\prod_{n=0}^{\infty} \left(1 - q^{2n+2j+2}\right) = \sum_{k=0}^{\infty} \frac{(-1)^k q^{k^2+2kj+k}}{(1-q^2)(1-q^4)\cdots(1-q^{2k})}.$$

Putting this into (11.3) yields

$$\prod_{n=1}^{\infty}(1 - q^{2n})(1 + q^{2n-1}z) = \sum_{j=-\infty}^{\infty} \sum_{k=0}^{\infty} \frac{(-1)^k q^{(j+k)^2+k} z^j}{(1-q^2)(1-q^4)\cdots(1-q^{2k})}$$

$$= \sum_{j=-\infty}^{\infty} \sum_{k=0}^{\infty} \frac{q^{j^2} z^j (-qz^{-1})^k}{(1-q^2)(1-q^4)\cdots(1-q^{2k})}, \quad (11.4)$$

where the last line is obtained by shifting the index j to $j - k$. Now we use E2 with q replaced by q^2 and then x by $-qz^{-1}$ to get

$$\prod_{n=1}^{\infty} \frac{1}{(1 + q^{2n-1}z^{-1})} = \sum_{k=0}^{\infty} \frac{(-qz^{-1})^k}{(1-q^2)(1-q^4)\cdots(1-q^{2k})}. \quad (11.5)$$

Therefore, from (11.4) and (11.5), we have

$$\prod_{n=1}^{\infty}(1 - q^{2n})(1 + q^{2n-1}z) = \sum_{j=-\infty}^{\infty} \left(q^{j^2} z^j \prod_{n=1}^{\infty} \frac{1}{(1 + q^{2n-1}z^{-1})}\right),$$

which is equivalent to (11.1). $\square$

12

Classical Partition Function and Euler's Product Formula

With various substitution of q and z, Jacobi's triple product identity gives many interesting results. For example, if we put $q = q^{3/2}$ and then $z = -q^{-1/2}$ into (11.1), we get

$$\sum_{n\in\mathbb{Z}}(-1)^n q^{\frac{\left(3n^2-n\right)}{2}} = \prod_{n=1}^{\infty}\left(1-q^{3n}\right)\left(1-q^{3n-2}\right)\left(1-q^{3n-1}\right) = \prod_{n=1}^{\infty}\left(1-q^n\right),$$
(12.1)

which is called *Euler's product formula*. We proved that it holds when $|q| < 1$. It follows that it also holds as an equality of formal power series in q (see Chapter 8). The formula may also be written using Euler's product

$$\varphi(q) = \prod_{n=1}^{\infty}(1-q^n)$$

as

$$\varphi(q) = \sum_{n\in\mathbb{Z}}(-1)^n q^{e_n},$$
(12.2)

where

$$e_n = \frac{3n^2-n}{2}$$
(12.3)

are called *pentagonal numbers*. The reader is encouraged to multiply out the first few factors of Euler's product to discover the astonishing fact that indeed the e_nth coefficient is $(-1)^n$ and all other coefficients are zero.

Definition. The *classical partition function* $p(n)$ is defined on the set of integers by letting $p(n)$ be the number of ways to partition n into a sum of positive integers (not counting the order of summands) if $n > 0$, $p(n) = 0$ if $n < 0$, and $p(0) = 1$.

For example, $p(1) = 1$ because the only way to write 1 as a sum is $1 = 1$, $p(2) = 2$ because $2 = 2 = 1 + 1$, $p(3) = 3$ because $3 = 3 = 2 + 1 = 1 + 1 + 1$, $p(4) = 5$ because $4 = 4 = 3 + 1 = 2 + 2 = 2 + 1 + 1 = 1 + 1 + 1 + 1$, and so on. This slow growth of $p(n)$ for small values of n is deceptive, since in fact, one knows that

$$p(n) \sim \frac{1}{4\sqrt{3}n} e^{\pi\sqrt{2n/3}} \text{ as } n \to \infty.$$

Thus the time needed to enumerate all partitions on n grows exponentially with n.

The following proposition shows how $\varphi(q)$ is related to the partitions of integers.

Proposition 12.1. *One has the following equality of formal power series in q:*

$$\frac{1}{\varphi(q)} = \sum_{n=0}^{\infty} p(n)q^n. \tag{12.4}$$

Proof. This well-known argument will often be used in the subsequent Chapters. Assuming that $|q| < 1$ and using the geometric series expansion, we have

$$
\begin{aligned}
\frac{1}{\varphi(q)} &= \frac{1}{(1-q)(1-q^2)(1-q^3)\cdots} \\
&= (1 + q + q^2 + q^3 + \cdots)(1 + q^2 + q^4 + q^6 + \cdots) \\
&\quad \times (1 + q^3 + q^6 + q^9 + \cdots) \cdots.
\end{aligned}
$$

Note that the exponents of q in the nth factor are all the nonnegative integer multiples of n. If we expand the product on the right-hand side into a power series, each term will be of the form $q^{n_1} q^{2n_2} q^{3n_3} \cdots = q^{1n_1 + 2n_2 + 3n_3 + \cdots}$, where n_i are all nonnegative integers. A q^n term is obtained if $n = 1n_1 + 2n_2 + 3n_3 + \cdots$, for some n_i, and each q^n term corresponds to a way to express n as a sum of positive integers, i.e., the sum of n_1 1's, n_2 2's, n_3 3's and so on. Also, a different way to partition n will contribute one new q^n term. Therefore, the coefficient of q^n, or the number of q^n terms, is exactly the number of ways to partition n into a sum of positive integers. $\square$

With the above interpretation of $\varphi(q)$, we can apply Euler's product identity to obtain a relation among the numbers $p(n)$. This relation is stated in the following theorem.

Theorem 12.1. *For any positive integer n we have*

$$
\begin{aligned}
p(n) \; = \; & p(n - e_1) + p(n - e_{-1}) - p(n - e_2) - p(n - e_{-2}) \\
& + p(n - e_3) + p(n - e_{-3}) + \cdots ,
\end{aligned}
\tag{12.5}
$$

where e_n are the pentagonal numbers defined by (12.3).

Proof. Using (12.2) and (12.4), we have

$$
1 = \left(\sum_{j \in \mathbb{Z}} (-1)^j q^{e_j} \right) \left(\sum_{k \in \mathbb{Z}} p(k) q^k \right).
\tag{12.6}
$$

When we expand the product, we obtain a $(-1)^j p(k) q^n$ term if $n = e_j + k$ for some integers j and k. Hence, for $n > 0$, we will obtain

$$
\begin{aligned}
0 \; = \; & p(n - e_0) - p(n - e_1) - p(n - e_{-1}) + p(n - e_2) + p(n - e_{-2}) \\
& - p(n - e_3) - p(n - e_{-3}) + \cdots ,
\end{aligned}
\tag{12.7}
$$

when equating the coefficients on both sides. Since $e_0 = 0$, the proof is complete. $\square$

Formula (12.5) is a very convenient recursive formula for a rapid calculation of $p(n)$. For example, $p(5) = p(4) + p(3) - p(0) = 5 + 3 - 1 = 7$, $p(6) = p(5) + p(4) - p(1) = 11$, etc. The time needed to evaluate $p(n)$ using formula (12.5) grows slower than n, which is, of course, much less than that required to enumerate all partitions of n.

As hinted by its name, the pentagonal numbers e_n have a geometrical meaning. This is described in the following picture:

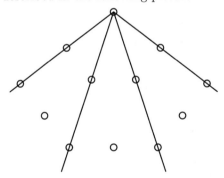

Pentagons that are similar to each other can be drawn by joining the nth vertices on each ray. The numbers of vertices enclosed by the pentagons (counting those on the edges) are the pentagonal numbers. For example, the first nontrivial pentagon encloses $e_2 = 5$ vertices and the second one $e_3 = 12$ vertices. The general formula for e_n under this interpretation can easily be proved by induction to be (12.3).

Now that we have pentagonal numbers, we can define m-gonal numbers for any $m \geq 3$ in a similar manner. The two most common types of polygonal numbers are the *triangular numbers*,

$$\Delta_n = \frac{n(n+1)}{2},$$

and, of course, the square numbers,

$$\square_n = n^2.$$

In general, we may deduce geometrically the formula for the nth m-gonal number:

$$m_n = (m-2)\Delta_{n-1} + n = \frac{n(mn - 2n - m + 4)}{2}.$$

Identities for triangular and square numbers similar to Euler's product formula can also be derived from Jacobi's triple product identity. Both of the identities given below were discovered by Gauss (before Jacobi discovered his triple product identity).

Proposition 12.2.

$$\sum_{n=0}^{\infty} q^{\Delta_n} = \prod_{n=1}^{\infty} \frac{1 - q^{2n}}{1 - q^{2n-1}}. \tag{12.8}$$

Proof. Substitute both q and z by $q^{1/2}$ in (11.1). We obtain

$$\sum_{n \in \mathbb{Z}} q^{\Delta_n} = \prod_{n=1}^{\infty} (1 - q^n)(1 + q^n)(1 + q^{n-1})$$

$$= 2 \prod_{n=1}^{\infty} (1 - q^{2n})(1 + q^n).$$

Observe that $\Delta_n = \Delta_{-n-1}$, and the sum from $n = 0$ to ∞ is thus the same as from $n = -1$ back to $-\infty$. Hence, we have

$$\sum_{n=0}^{\infty} q^{\Delta_n} = \prod_{n=1}^{\infty} (1 - q^{2n})(1 + q^n). \tag{12.9}$$

Since

$$\prod_{n=1}^{\infty} (1 + q^n) = \prod_{n=1}^{\infty} \frac{1 - q^{2n}}{1 - q^n} = \prod_{n=1}^{\infty} \frac{1}{1 - q^{2n-1}}, \tag{12.10}$$

the desired result is obtained. $\square$

Proposition 12.3.

$$\sum_{n \in \mathbb{Z}} (-q)^{n^2} = \prod_{n=1}^{\infty} \frac{1 - q^n}{1 + q^n}. \tag{12.11}$$

Proof. If we put $z = -1$ in (11.1), we have

$$\sum_{n\in\mathbb{Z}}(-q)^{n^2} = \prod_{n=1}^{\infty}(1 - q^{2n})(1 - q^{2n-1})(1 - q^{2n-1})$$

$$= \prod_{n=1}^{\infty}(1 - q^n)(1 - q^{2n-1}) = \prod_{n=1}^{\infty}\frac{1 - q^n}{1 + q^n},$$

where we have used (12.10) in the last equality. □

Identities (12.8) and (12.11) will be useful later when we study the partition of an integer as a sum of triangular numbers or of square numbers. Before going on, the combinatorial meaning of (12.10) deserves a remark. The product on the left in (12.10) is

$$(1 + q)(1 + q^2)(1 + q^3)\cdots .$$

A q^n term appears in the expansion if $n = a_1 + a_2 + a_3 + \cdots$, where a_i are distinct. Using the similar argument in the proof of Proposition 12.1, the coefficient of q^n is the number of ways to write n as a sum of *distinct* positive integers. On the other hand, the product on the other side is

$$(1 + q + q^2 + q^3 + \cdots)(1 + q^3 + q^6 + q^9 + \cdots)(1 + q^5 + q^{10} + q^{15} + \cdots)\cdots .$$

Each q^n term corresponds to a way of expressing n as a sum of odd numbers. Therefore, (12.10) says that the number of ways to partition n into distinct positive numbers is the same as the number of ways to partition n into odd numbers.

To conclude this chapter, we briefly interrupt our discussion of q-calculus by introducing an important function in number theory that shares with $p(n)$ the same recursive relation (12.5).

Theorem 12.2. *For any nonzero integer n, define a function*

$$d(n) = \begin{cases} \text{sum of positive divisors of } n, & n > 0, \\ 0 & n < 0. \end{cases}$$

Then, for $n > 0$ we have

$$d(n) = d(n - e_1) + d(n - e_{-1}) - d(n - e_2) - d(n - e_{-2}) + \cdots , \quad (12.12)$$

where we take $d(0) = n$ if it enters the right-hand side.

Proof. Define the generating function

$$D(q) = \sum_{n=1}^{\infty} d(n)q^n.$$

Switching the order of summation, we have

$$D(q) \;=\; \sum_{n=1}^{\infty}\sum_{m|n} mq^n = \sum_{m=1}^{\infty}\sum_{m|n} mq^n$$

$$=\; \sum_{m=1}^{\infty}\sum_{m|n} m(q^m + q^{2m} + q^{3m} + \cdots) = \sum_{m=1}^{\infty}\sum_{m|n} \frac{mq^m}{1-q^m}$$

$$=\; -q\sum_{m=1}^{\infty}\frac{d}{dq}\log(1-q^m) = -q\frac{d}{dq}\log\prod_{m=1}^{\infty}(1-q^m)$$

$$=\; \frac{-q\frac{d}{dq}\prod_{m=1}^{\infty}(1-q^m)}{\prod_{m=1}^{\infty}(1-q^m)},$$

or

$$D(q)\varphi(q) = -q\frac{d}{dq}\varphi(q).$$

Using (12.2), we have

$$\left(\sum_{j=1}^{\infty} d(j)q^j\right)\left(\sum_{k\in\mathbb{Z}}(-1)^k q^{e_k}\right) = \sum_{m\in\mathbb{Z}}(-1)^{m+1}e_m q^{e_m}.$$

Comparing the coefficients of q^n on both sides, we have

$$\sum_{k\in\mathbb{Z}}(-1)^k d(n - e_k) = \begin{cases} (-1)^{m+1}e_m & \text{if } n = e_m \text{ for some } m \in \mathbb{Z}, \\ 0 & \text{otherwise,} \end{cases}$$

which is the same as (12.12), since $j = n - e_k \geq 1$ and $d(n)$ is defined to be zero for any negative n. $\square$

13

q-Hypergeometric Functions and Heine's Formula

For further study of infinite sums and infinite products we would like to introduce the *hypergeometric series*. A classical hypergeometric series is defined as follows.

Definition. $F(x)$ is a hypergeometric series if

$$F(x) = \sum_{n=0}^{\infty} c_n x^n, \tag{13.1}$$

where

$$\frac{c_{n+1}}{c_n} = R(n), \qquad c_0 = 1, \tag{13.2}$$

and R is a rational function whose denominator does not vanish at nonnegative integers.

If $R(t)$ is given, the coefficients c_n are immediately determined:

$$c_n = R(0)R(1)\cdots R(n-1).$$

For example, if $R(t) \equiv 1$, then $c_n = 1$ for all n, and $F(x)$ is a geometric series. Rescaling t if necessary, we may factorize $R(t)$ into the following form, up to a constant factor:

$$R(t) = \frac{(t+a_1)(t+a_2)\cdots(t+a_r)}{(t+b_1)(t+b_2)\cdots(t+b_s)(t+1)}, \tag{13.3}$$

where $a_i \neq b_j$, and b_j are not nonpositive integers, for all $1 \leq i \leq r$, $1 \leq j \leq s$.

A notation introduced by Gauss summarizes the essential information of a general hypergeometric series. If $F(x)$ is as defined by (13.1) and (13.2), and $R(t)$ has the form (13.3), one writes

$$F(x) = {}_rF_s\left[\begin{array}{c} a_1, \ldots, a_r \\ b_1, \ldots, b_s \end{array}; x\right], \tag{13.4}$$

which, when expressed explicitly, is

$$1 + \sum_{n=1}^{\infty} \frac{\{a_1(a_1+1)\cdots(a_1+n-1)\}\cdots\{a_r(a_r+1)\cdots(a_r+n-1)\}}{\{b_1(b_1+1)\cdots(b_1+n-1)\}\cdots\{b_s(b_s+1)\cdots(b_s+n-1)\}}\frac{x^n}{n!}. \tag{13.5}$$

For example,

$$_0F_0[x] = 1 + \sum_{n=1}^{\infty} \frac{x^n}{n!} = e^x,$$

and

$$_1F_0\left[\begin{array}{c} a \\ - \end{array}; x\right] = 1 + \sum_{n=1}^{\infty} \frac{a(a+1)\cdots(a+n-1)}{n!}x^n = \frac{1}{(1-x)^a}.$$

Thus, the hypergeometric series is a general type of series, for which many series, like geometric, binomial, and exponential series, are special cases.

The q-analogue of hypergeometric series was first introduced by Heine.

Definition. $\Phi(x)$ is a q-hypergeometric series if

$$\Phi(x) = \sum_{n=0}^{\infty} c_n x^n, \tag{13.6}$$

where

$$\frac{c_{n+1}}{c_n} = R(q^n), \quad c_0 = 1, \tag{13.7}$$

and $R(t)$ is a rational function whose denominator does not vanish at $t = 1, q, q^2, \ldots$.

Similarly, we have, for $n \geq 1$,

$$c_n = R(1)R(q)\cdots R\left(q^{n-1}\right). \tag{13.8}$$

By convention, the rational function R is considered in a slightly different form:

$$R(t) = \frac{\left(a_1 - t^{-1}\right)\cdots\left(a_r - t^{-1}\right)}{\left(b_1 - t^{-1}\right)\cdots\left(b_s - t^{-1}\right)\left(q - t^{-1}\right)}. \tag{13.9}$$

Here, $a_i \neq b_j$ and each b_j is not one of $1, q^{-1}, q^{-2}, \ldots$. Then, since

$$\prod_{j=0}^{n-1}(a - q^{-j}) = \prod_{j=0}^{n-1}(-q^{-j})(1 - q^j a) = (-1)^n q^{-n(n-1)/2}(1-a)_q^n, \tag{13.10}$$

we have

$$c_n = \left\{(-1)^n q^{n(n-1)/2}\right\}^{s-r+1} \frac{(1-a_1)_q^n \cdots (1-a_r)_q^n}{(1-b_1)_q^n \cdots (1-b_s)_q^n} \frac{1}{(1-q)_q^n}. \quad (13.11)$$

The q-hypergeometric series also has a notation similar to (13.4):

$$\Phi(x) = {}_r\Phi_s\left[\begin{array}{c} a_1,\ldots,a_r \\ b_1,\ldots,b_s \end{array}; q; x\right]. \quad (13.12)$$

For example, from (9.3), (9.4), (9.8), and (9.10) we have

$$_0\Phi_0\left[\begin{array}{c} - \\ - \end{array}; q; x\right] = \sum_{n=0}^{\infty} \frac{(-1)^n q^{\frac{n(n-1)}{2}}}{(1-q)_q^n} x^n = (1-x)_q^{\infty} = E_q^{x/(q-1)} \quad (13.13)$$

and

$$_1\Phi_0\left[\begin{array}{c} 0 \\ - \end{array}; q; x\right] = \sum_{n=0}^{\infty} \frac{1}{(1-q)_q^n} x^n = \frac{1}{(1-x)_q^{\infty}} = e_q^{x/(1-q)}. \quad (13.14)$$

Let us now examine the next-simplest q-hypergeometric series, i.e.,

$$_1\Phi_0[a; q; x] = 1 + \sum_{n=1}^{\infty} \frac{(1-a)_q^n}{(1-q)_q^n} x^n. \quad (13.15)$$

Here we write $_1\Phi_0\left[\begin{array}{c} a \\ - \end{array}; q; x\right]$ as $_1\Phi_0[a; q; x]$ for simplicity. If $a = q^N$, where N is a positive integer, we have

$$
\begin{aligned}
_1\Phi_0[q^N; q; x] &= 1 + \sum_{n=1}^{\infty} \frac{(1-q^N)\cdots(1-q^{N+n-1})}{(1-q)\cdots(1-q^{n-1})} x^n \\
&= 1 + \sum_{n=1}^{\infty} \frac{[N]\cdots[N+n-1]}{[n]!} x^n.
\end{aligned}
$$

According to Heine's binomial formula (8.1), we have

$$_1\Phi_0[q^N; q; x] = \frac{1}{(1-x)_q^N}. \quad (13.16)$$

This result inspires us with the following theorem.

Theorem 13.1. *For any a, one has the following formula of Heine's:*

$$_1\Phi_0[a; q; x] = \frac{(1-ax)_q^{\infty}}{(1-x)_q^{\infty}}. \quad (13.17)$$

Proof. Firstly, (13.17) is true if $a = q^N$, because

$$\frac{(1-q^N x)_q^{\infty}}{(1-x)_q^{\infty}} = \frac{(1-q^N x)(1-q^{N+1}x)\cdots}{(1-x)(1-qx)\cdots} = \frac{1}{(1-x)_q^N},$$

which is equal to $_1\Phi_0[a; q; x]$ by (13.16).

To complete the proof, we apply a very useful argument. Now, both sides in (13.17) can be expressed as infinite series with coefficients being rational functions of a, i.e.,

$$_1\Phi_0[a; q; x] = \sum_{n=0}^{\infty} c_n(a)x^n, \quad \frac{(1 - ax)_q^{\infty}}{(1 - x)_q^{\infty}} = \sum_{n=0}^{\infty} c_n'(a)x^n.$$

We know from above that for each n, $c_n = c_n'$ at infinitely many different values of a, namely, $a = q^N$, where N is a positive integer. In other words, $c_n - c_n'$ is a rational function of a with infinitely many zeros. Such a rational function must be identically zero, because the number of zeros of a rational function cannot exceed the degree of the polynomial in its numerator. Therefore, $c_n = c_n'$ for each n, and the proof is complete.   □

14
More on Heine's Formula and the General Binomial

Inspired by (13.16) and (13.17), it is natural to generalize the notion of a q-binomial in the following way.

Definition. For any number α, define

$$(1+x)_q^\alpha = \frac{(1+x)_q^\infty}{(1+q^\alpha x)_q^\infty}. \tag{14.1}$$

Obviously, this definition coincides with the original one given by (3.4) when α is a positive integer, and also with the one given by (3.7) when α is a negative integer. The fact that it is an appropriate generalization is justified by the following two propositions, which are generalizations of Proposition 3.2 and equation (3.11).

Proposition 14.1. *For any two numbers α and β, we have*

$$(1+x)_q^\alpha (1+q^\alpha x)_q^\beta = (1+x)_q^{\alpha+\beta}. \tag{14.2}$$

Proof. The proposition follows directly from the definition, since

$$\frac{(1+x)_q^\infty}{(1+q^\alpha x)_q^\infty} \frac{(1+q^\alpha x)_q^\infty}{(1+q^{\alpha+\beta} x)_q^\infty} = \frac{(1+x)_q^\infty}{(1+q^{\alpha+\beta} x)_q^\infty}. \quad \square$$

Proposition 14.2. *For any number α, we have*

$$D_q(1+x)_q^\alpha = [\alpha](1+qx)_q^{\alpha-1}. \tag{14.3}$$

Proof. By definition, we have

$$
D_q \left(\frac{(1+x)_q^\infty}{(1+q^\alpha x)_q^\infty} \right) = \left(\frac{(1+qx)_q^\infty}{(1+q^{\alpha+1}x)_q^\infty} - \frac{(1+x)_q^\infty}{(1+q^\alpha x)_q^\infty} \right) \frac{1}{(q-1)x}
$$

$$
= \frac{(1+qx)_q^\infty}{(1+q^\alpha x)_q^\infty} \frac{(1+q^\alpha x) - (1+x)}{(q-1)x}
$$

$$
= (1+qx)_q^{\alpha-1} \frac{q^\alpha - 1}{q-1}.
$$

With the definition of $[\alpha]$ given by (3.8), the proof is complete. $\square$

Proposition 14.2 allows us to compute the Taylor series of $(1+x)_q^\alpha$. Using the chain rule (1.15), we obtain

$$
\begin{aligned}
D_q{}^j (1+x)_q^\alpha &= D_q{}^{j-1}[\alpha](1+qx)_q^{\alpha-1} \\
&= D_q{}^{j-2}[\alpha] \cdot q[\alpha - 1](1+q^2 x)_q^{\alpha-2} \\
&= D_q{}^{j-3}[\alpha] \cdot q[\alpha - 1] \cdot q^2[\alpha - 2](1+q^3 x)_q^{\alpha-3} \\
&\quad \vdots \\
&= [\alpha] \cdot q[\alpha - 1] \cdot q^2[\alpha - 2] \cdots q^{j-1}[\alpha - j + 1](1+q^j x)_q^{\alpha-j},
\end{aligned}
$$

and thus $D_q{}^j (1+x)_q^\alpha \big|_{x=0} = q^{j(j-1)/2}[\alpha][\alpha - 1] \cdots [\alpha - j + 1]$. Therefore, we have

$$
(1+x)_q^\alpha = \sum_{j=0}^\infty \frac{q^{j(j-1)/2}[\alpha][\alpha - 1] \cdots [\alpha - j + 1]x^j}{[j]!}
$$

$$
= \sum_{j=0}^\infty \begin{bmatrix} \alpha \\ j \end{bmatrix} q^{j(j-1)/2} x^j, \tag{14.4}
$$

which generalizes Gauss's binomial formula (5.5).

Proposition 14.3. *For any number α, we have*

$$
D_q \left(\frac{1}{(1-x)_q^\alpha} \right) = \frac{[\alpha]}{(1-x)_q^{\alpha+1}}. \tag{14.5}
$$

Proof. By the definition of q-derivative, we have

$$
D_q \left(\frac{1}{(1-x)_q^\alpha} \right) = D_q \left(\frac{(1-q^\alpha x)_q^\infty}{(1-x)_q^\infty} \right)
$$

$$
= \left(\frac{(1-q^{\alpha+1}x)_q^\infty}{(1-qx)_q^\infty} - \frac{(1-q^\alpha x)_q^\infty}{(1-x)_q^\infty} \right) \frac{1}{(q-1)x}
$$

$$
= \frac{(1-q^{\alpha+1}x)_q^\infty}{(1-x)_q^\infty} \frac{(1-x) - (1-q^\alpha x)}{(q-1)x}
$$

$$
= \frac{1}{(1-x)_q^{\alpha+1}} \frac{q^\alpha - 1}{q-1},
$$

as desired. $\square$

Using this proposition and induction, one can easily see that

$$D_q{}^j \left(\frac{1}{(1-x)_q^\alpha} \right) \Bigg|_{x=0} = [\alpha][\alpha+1]\cdots[\alpha+j-1]. \qquad (14.6)$$

Hence, we have the Taylor expansion

$$\frac{1}{(1-x)_q^\alpha} = \sum_{j=0}^\infty \frac{[\alpha][\alpha+1]\cdots[\alpha+j-1]x^j}{[j]!}$$

$$= \sum_{j=0}^\infty \frac{(1-q^\alpha)(1-q^{\alpha+1})\cdots(1-q^{\alpha+j-1})x^j}{(1-q)(1-q^2)\cdots(1-q^j)}$$

$$= \sum_{j=0}^\infty \frac{(1-q^\alpha)_q^j \, x^j}{(1-q)_q^j}.$$

By (13.15) and the recognition of q^α as a, we recover Heine's formula (13.17).

15
Ramanujan Product Formula

In this chapter, we apply Heine's formula to prove a remarkable identity discovered by the Indian mathematician Ramanujan. This identity relates a *bilateral q-hypergeometric series* to an infinite product, and it has many interesting applications in number theory, which will be discussed in subsequent chapters.

In order to prove Ramanujan's formula we shall need some elementary facts from the theory of complex analytic functions. A formal power series in z that converges in the open disk $D_\epsilon = \{z : |z| < \epsilon\}$ on the complex plane for some $\epsilon > 0$ is called an analytic function (in D_ϵ). Of course, all polynomials in z with arbitrary complex coefficients are analytic functions (in D_∞). A less obvious example of an analytic function in D_∞ is $(1+z)_q^\infty$. This follows from (9.3) by applying the ratio test. It is easy to show that any linear combination and product of analytic functions is an analytic function; also, if $f(z)$ is analytic, then $1/f(z)$ is analytic, provided that $f(z)$ has no zero in D_ϵ.

A series $\sum_{n=1}^\infty f_n(z)$ of analytic functions in D_ϵ converges to an analytic function in D_ϵ if for each n, $|f_n(z)| \leq M_n$ on D_ϵ for some M_n such that $\sum_n M_n$ converges. In fact, the coefficients in the series expansion $f_n(z) = \sum_{m=0}^\infty a_{nm} z^m$ may be expressed as

$$a_{nm} = \frac{1}{2\pi i} \oint \frac{f_n(z)}{z^{m+1}} dz, \qquad n \geq 1, \, m \geq 0,$$

where the integral is evaluated along a circle centered at the origin with radius r, $0 < r < \epsilon$. (Those who have not encountered this formula may easily prove it by expressing $f_n(z)$ in its series form and making the substitution $z = re^{i\theta}$, $0 < \theta < 2\pi$.) One may easily deduce from the formula that $|a_{nm}| \leq M_n r^{-m}$, or, since r can be arbitrarily close to ϵ, $|a_{nm}| \leq M_n \epsilon^{-m}$. For each m we define $a_m = \sum_n a_{nm}$, which is a convergent series because $|a_m| \leq \sum_n |a_{nm}| \leq M\epsilon^{-m}$ where $M = \sum_n M_n$. Then, the formal power series $f(z) = \sum_{m=0}^{\infty} a_m z^m$ actually defines an analytic function in D_ϵ, since

$$|f(z)| \leq \sum_{m=0}^{\infty} |a_m| |z|^m \leq M \sum_{m=0}^{\infty} \left(\frac{|z|}{\epsilon}\right)^m,$$

and the latter series converges for any $|z| < \epsilon$. The same estimate $|a_m| < M\epsilon^{-m}$ shows that $f(z) = \sum_{n=1}^{\infty} f_n(z)$ converges to $f(z)$ in D_ϵ.

Analytic functions have a property similar to that of polynomials: If an analytic function $f(z)$ has an infinite number of roots in D_ϵ, say $\{z_1, z_2, \ldots\}$, such that $\lim_{i\to\infty} z_i = 0$, then $f(z)$ is identically zero. Indeed, in the contrary case we can write $f(z) = \sum_{j\geq n} c_j z^j$, where $c_n \neq 0$. Since $f(z_i) = \sum_{j\geq n} c_j z_i^j = 0$, dividing by z_i^n, we obtain for any $i \geq 1$,

$$c_n + c_{n+1} z_i + c_{n+2} z_i^2 + \cdots = 0.$$

Taking $i \to \infty$, we obtain $c_n = 0$, a contradiction. (The requirement of $\lim_{i\to\infty} z_i = 0$ is crucial, since there exist nonzero analytic functions possessing infinitely many roots, e.g., all roots of $e^z - 1$ are $2\pi in$, where n is any integer.)

Let us now state Ramanujan's formula.

Theorem 15.1. *(Ramanujan product formula) In the domain*

$$|q| < 1, \quad |a| > |q|, \quad |b| < 1, \quad and \quad \left|\frac{b}{a}\right| < |x| < 1, \tag{15.1}$$

we have the following equality of functions in a, b, q, x:

$$
\begin{aligned}
{}_1\Psi_1(a, b; q; x) : \quad &= \quad \sum_{n=-\infty}^{\infty} \frac{(1-a)_q^n}{(1-b)_q^n} x^n \\
&= \quad \prod_{n=0}^{\infty} \frac{(1 - \frac{bq^n}{a})(1 - q^{n+1})(1 - \frac{q^{n+1}}{ax})(1 - axq^n)}{(1 - bq^n)(1 - \frac{q^{n+1}}{a})(1 - \frac{bq^n}{ax})(1 - xq^n)}. \tag{15.2}
\end{aligned}
$$

Proof (M.E.H. Ismail). The strategy is first to show that both sides of (15.2) are analytic functions in b (when a, q and x are fixed) in the nonempty domain specified by (15.1) (i.e., $|b| < \min\{1, |ax|\}$), and then to prove the equality when $b = q^M, q^{M+1}, q^{M+2}, \ldots$, M being a sufficiently large integer. Hence, the difference of the two sides of (15.2) is an analytic function in b possessing an infinite sequence of roots that converges to zero, and must therefore vanish identically.

Denote by $f_n(b)$ the nth term in the series in (15.2). To see that the series converges to an analytic function in b, it suffices as noted above, to show that $f_n(b)$ is analytic for any $n \in \mathbb{Z}$ and that as $n \to \pm\infty$, $|f_n(b)| < M_\pm c_\pm^{|n|}$ for some $0 < c_\pm < 1$ and positive constants $M_\pm$. If $n > 0$, then $(1 - b)_q^n$ is a polynomial in b that does not vanish in the domain, because $|b|, |q| < 1$, thus showing the analyticity of $f_n(b)$. Since as $n \to \infty$,

$$\left| \frac{f_{n+1}(b)}{f_n(b)} \right| = \left| \frac{(1 - q^n a)x}{1 - q^n b} \right| \longrightarrow |x| < 1,$$

we have $|f_{n+1}(b)| < c_+|f_n(b)|$ for any $c_+ \in (|x|, 1)$ when n is large enough, which implies that $|f_n(b)| < M_+ c_+^n$ for some M_+ and all n large enough. Similarly, for $n < 0$ we have by (3.7) that $1/(1 - b)_q^n = (1 - q^n b)_q^{|n|}$ is a polynomial, and thus analytic. We have

$$\left| \frac{f_{-n-1}(b)}{f_{-n}(b)} \right| = \left| \frac{1 - q^{-n-1}b}{(1 - q^{-n-1}a)x} \right| \longrightarrow \left| \frac{b}{ax} \right| < 1$$

as $n \to -\infty$, and the argument above may be applied again.

For the product side, we note that the factors in the denominator never vanish in the specified domain and each factor is either independent of b or is of the form $(1 + cb)_q^\infty$, where c is independent of b. As remarked before, $f(z) = (1 + z)_q^\infty$ is analytic in D_∞. Since products of analytic functions and reciprocals of nonvanishing analytic functions are also analytic, the fraction above is analytic in the specified domain.

We may now perform the substitution $b = q^M$, where M is so large that b lies within the specified domain. Noting that

$$\frac{1}{(1 - q^M)_q^n} = \frac{1}{(1 - q^M)_q^{-n'}} = (1 - q^{M-n'})_q^{n'} = 0$$

if $n = -n' \leq -M$, we see the left-hand side (LHS) of (15.2) become

$$\sum_{n \in \mathbb{Z}} \frac{(1 - a)_q^n}{(1 - q^M)_q^n} x^n = \sum_{n=-M+1}^{\infty} \frac{(1 - a)_q^n}{(1 - q^M)_q^n} x^n$$

$$= x^{1-M} \sum_{n=0}^{\infty} \frac{(1 - a)_q^{n+1-M}}{(1 - q^M)_q^{n+1-M}} x^n$$

$$= x^{1-M} \frac{(1 - a)_q^{1-M}}{(1 - q^M)_q^{1-M}} \sum_{n=0}^{\infty} \frac{(1 - q^{1-M}a)_q^n}{(1 - q)_q^n} x^n$$

$$= x^{1-M} \frac{(1 - a)_q^{1-M}}{(1 - q^M)_q^{1-M}} {}_1\Phi_0[q^{1-M}a; q; x],$$

where we have applied (14.2) with $\alpha = 1 - M$ and $\beta = n$. By Heine's formula (13.17), we have

$$
\begin{aligned}
\text{LHS} &= x^{1-M} \frac{(1-a)_q^{1-M}}{(1-q^M)_q^{1-M}} \frac{(1-q^{1-M}ax)_q^\infty}{(1-x)_q^\infty} \\
&= x^{1-M} \frac{(1-q)_q^{M-1}}{(1-q^{1-M}a)_q^{M-1}} \frac{(1-q^{1-M}ax)_q^\infty}{(1-x)_q^\infty},
\end{aligned}
\tag{15.3}
$$

where (3.7) is used in the last equality. To complete the proof, we begin comparing the two sides of (15.2). Putting $b = q^M$ in the product side, we have

$$
\begin{aligned}
\text{RHS} &= \frac{(1-q^M a^{-1})_q^\infty (1-q)_q^\infty (1-qa^{-1}x^{-1})_q^\infty (1-ax)_q^\infty}{(1-q^M)_q^\infty (1-qa^{-1})_q^\infty (1-q^M a^{-1}x^{-1})_q^\infty (1-x)_q^\infty} \\
&= \frac{(1-q)_q^{M-1}}{(1-qa^{-1})_q^{M-1}} \frac{(1-qa^{-1}x^{-1})_q^{M-1}(1-ax)_q^\infty}{(1-x)_q^\infty}.
\end{aligned}
\tag{15.4}
$$

Comparing (15.3) and (15.4), we see that what is left to show is

$$
x^{1-M} \frac{(1-q^{1-M}ax)_q^\infty}{(1-q^{1-M}a)_q^{M-1}} = \frac{(1-qa^{-1}x^{-1})_q^{M-1}(1-ax)_q^\infty}{(1-qa^{-1})_q^{M-1}},
$$

or, equivalently,

$$
\begin{aligned}
\frac{(1-qa^{-1})_q^{M-1}}{(1-q^{1-M}a)_q^{M-1}} &= \frac{(1-qa^{-1}x^{-1})_q^{M-1}(1-ax)_q^\infty}{(1-q^{1-M}ax)_q^\infty} x^{M-1} \\
&= \frac{(1-qa^{-1}x^{-1})_q^{M-1}}{(1-q^{1-M}ax)_q^{M-1}} x^{M-1}.
\end{aligned}
$$

Observe that the left side is independent of x, and the two sides agree when $x = 1$. Therefore, we hope that the right-hand side is also independent of x. Expanding the right-hand side, we have

$$
\begin{aligned}
&\frac{(1-qa^{-1}x^{-1})(1-q^2a^{-1}x^{-1})\cdots(1-q^{M-1}a^{-1}x^{-1})x^{M-1}}{(1-q^{1-M}ax)(1-q^{2-M}ax)\cdots(1-q^{-1}ax)} \\
&= \frac{1-qa^{-1}x^{-1}}{1-q^{-1}ax} \frac{1-q^2a^{-1}x^{-1}}{1-q^{-2}ax} \cdots \frac{1-q^{M-1}a^{-1}x^{-1}}{1-q^{1-M}ax} x^{M-1} \\
&= \left(-\frac{q}{ax}\right)\left(-\frac{q^2}{ax}\right)\cdots\left(-\frac{q^{M-1}}{ax}\right) x^{M-1},
\end{aligned}
$$

which is independent of x, as desired. $\square$

The Ramanujan product identity involves four variables, q, a, b, and x. One of the important special cases is $a = y$, $b = qy$. The domain of convergence becomes

$$
|q| < |x| < 1 \quad \text{and} \quad |q| < |y| < |q|^{-1},
\tag{15.5}
$$

which is nonempty. The left-hand side of (15.2) becomes

$$\text{LHS} \;=\; \sum_{n\in\mathbb{Z}} \frac{(1-y)_q^n}{(1-qy)_q^n} x^n = \sum_{n\in\mathbb{Z}} \frac{1-y}{1-q^n y} x^n,$$

because $(1-y)_q^n (1 - q^n y) = (1-y)_q^{n+1} = (1-y)(1-qy)_q^n$ for any integer n. If we restrict the domain of y to $|q| < |y| < 1$, we have $|q^n y| < 1$ if $n \geq 0$ and $|q^n y^{-1}| < 1$ if $n \geq 1$, and we can use a geometric series expansion to obtain

$$\text{LHS} \;=\; \sum_{n=0}^{\infty} \frac{(1-y)x^n}{1-q^n y} + \sum_{n=-1}^{-\infty} \frac{(1-y)x^n}{1-q^n y}$$

$$=\; \sum_{n=0}^{\infty} \frac{(1-y)x^n}{1-q^n y} - \sum_{n=1}^{\infty} \frac{(1-y)x^{-n}q^n y^{-1}}{1-q^n y^{-1}}$$

$$=\; (1-y)\left(\sum_{n=0}^{\infty}\sum_{m=0}^{\infty} x^n (q^n y)^m - \sum_{n=1}^{\infty}\sum_{m=1}^{\infty} x^{-n}(q^n y^{-1})^m \right).$$

On the other hand, the product side becomes

$$\prod_{n=0}^{\infty} \frac{(1-q^{n+1})^2 (1-x^{-1}y^{-1}q^{n+1})(1-xyq^n)}{(1-yq^{n+1})(1-x^{-1}q^{n+1})(1-y^{-1}q^{n+1})(1-xq^n)}.$$

Dividing both sides by $1-y$, we have

$$\sum_{m,n=0}^{\infty} q^{mn} x^n y^m - \sum_{m,n=1}^{\infty} q^{mn} x^{-n} y^{-m}$$

$$=\; \prod_{n=1}^{\infty} \frac{(1-q^n)^2 (1-x^{-1}y^{-1}q^n)(1-xyq^{n-1})}{(1-xq^{n-1})(1-x^{-1}q^n)(1-yq^{n-1})(1-y^{-1}q^n)}. \tag{15.6}$$

Hence, in the domain $|q| < |z| < 1$ we have

$$\sum_{m,n=0}^{\infty} q^{mn} z^{m+n} - \sum_{m,n=1}^{\infty} q^{mn} z^{-m-n}$$

$$=\; \prod_{n=1}^{\infty} \frac{(1-q^n)^2 (1-z^{-2}q^n)(1-z^2 q^{n-1})}{(1-zq^{n-1})^2 (1-z^{-1}q^n)^2} \tag{15.7}$$

if we put $x = y = z$. We can go further to pull the terms with $m = 0$ or $n = 0$ out from the first summation. Since the sum of these terms is

$$\sum_{n=0}^{\infty} z^n + \sum_{m=0}^{\infty} z^m - 1 = \frac{1+z}{1-z},$$

if we multiply both sides of (15.7) by $\frac{1-z}{1+z} = \frac{(1-z)^2}{1-z^2}$, we have in the same domain

$$1 + \frac{1-z}{1+z} \sum_{m,n=1}^{\infty} q^{mn}(z^{m+n} - z^{-m-n})$$

$$= \prod_{n=1}^{\infty} \frac{(1-q^n)^2(1-z^{-2}q^n)(1-z^2 q^n)}{(1-zq^n)^2(1-z^{-1}q^n)^2}. \tag{15.8}$$

The results above are important in our discussion of number-theoretic applications in the next two chapters.

16

Explicit Formulas for Sums of Two and of Four Squares

One of the oldest problems in number theory concerns the partition of an integer into a sum of squares. A famous result, first proved by Lagrange, is that any positive integer is a sum of four squares. In this chapter, we will not only prove this theorem, but also will find explicit formulas of Gauss and of Jacobi for the number of partitions of an integer into a sum of two and of four squares.

First, let us denote the number of ways to express N as a sum of m integer squares, counting the order, by $\Box_m(N)$. For example, $\Box_2(5) = 8$, because a total of eight ordered pairs, $(\pm 1, \pm 2)$, $(\pm 2, \pm 1)$, have their sum of squares equal to 5. If we define the formal power series

$$\Box(q) = \sum_{n \in \mathbb{Z}} q^{n^2}, \tag{16.1}$$

then we have

$$\Box_m(N) = \text{coefficient of } q^N \text{ in } \Box(q)^m. \tag{16.2}$$

To understand this, imagine that we expand the power series in (16.1):

$$\Box(q)^m = (\cdots + q^9 + q^4 + q + 1 + q + q^4 + q^9 + \cdots)^m.$$

In the resulting series, each q^N term corresponds in a one-to-one manner to an m-tuple $(a_1, \ldots, a_m)$, with $N = a_1^2 + \cdots + a_m^2$. Thus, the number of q^N appearing is the number of ways to express N as a sum of squares of m integers. For $m = 4$, we have the following theorem.

Theorem 16.1. *For any positive integer N we have*

$$\square_4(N) = 8 \times (\text{sum of positive divisors of } N \text{ that are not multiples of } 4).$$
$$(16.3)$$

An immediate corollary of the theorem is that any positive integer is a sum of four squares, since 1 is a divisor of any integer and 1 is not a multiple of 4. For example, $6 = 2^2 + 1^2 + 1^2 + 0^2$, $97 = 8^2 + 5^2 + 2^2 + 2^2$.

Proof of Theorem 16.1. Consider (15.8) and let $z \to -1$ on both sides. From (12.11), we have

$$\lim_{z \to -1} \text{RHS} = \left(\prod_{n=1}^{\infty} \frac{1 - q^n}{1 + q^n} \right)^4 = \square(-q)^4.$$

Writing the left-hand side as

$$1 + (1 - z) \sum_{m,n=1}^{\infty} q^{mn} \left(\frac{z^{m+n} - z^{-m-n}}{1 + z} \right)$$

and applying L'Hospital's rule, we have

$$\lim_{z \to -1} \text{LHS} = 1 + 4 \sum_{m,n=1}^{\infty} (-1)^{m+n-1}(m + n)q^{mn}.$$

By symmetry,

$$\sum_{m,n=1}^{\infty} (-1)^{m+n-1}mq^{mn} = \sum_{m,n=1}^{\infty} (-1)^{m+n-1}nq^{mn},$$

so we may rewrite the left-hand side as

$$\begin{aligned}
\text{LHS} &= 1 + 8 \sum_{m,n=1}^{\infty} (-1)^{m-1}m(-q^m)^n \\
&= 1 + 8 \sum_{m=1}^{\infty} \frac{(-1)^m mq^m}{1 + q^m} = 1 + 8 \sum_{m=1}^{\infty} \frac{m(-q)^m}{1 + q^m}.
\end{aligned}$$

Combining the results for the two sides and replacing q by $-q$, we have

$$\square(q)^4 = 1 + 8 \sum_{m=1}^{\infty} \frac{mq^m}{1 + (-q)^m}.$$
$$(16.4)$$

Since

$$\sum_{m=1}^{\infty} \frac{mq^m}{1 + (-q)^m} = \sum_{\substack{m \geq 1 \\ m \text{ odd}}} \frac{mq^m}{1 - q^m} + \sum_{\substack{m \geq 1 \\ m \text{ even}}} \frac{mq^m}{1 + q^m}$$

and

$$\sum_{k=1}^{\infty} \frac{2kq^{2k}}{1+q^{2k}} = \sum_{k=1}^{\infty} 2k(q^{2k} - q^{4k} + q^{6k} - q^{8k} + \cdots)$$

$$= \sum_{k=1}^{\infty} 2k(q^{2k} + q^{4k} + q^{6k} + q^{8k} + \cdots)$$

$$- \sum_{k=1}^{\infty} 4k(q^{4k} + q^{8k} + \cdots)$$

$$= \sum_{k=1}^{\infty} \frac{2kq^{2k}}{1-q^{2k}} - \sum_{k=1}^{\infty} \frac{4kq^{4k}}{1-q^{4k}},$$

we have

$$\sum_{m=1}^{\infty} \frac{mq^m}{1+(-q)^m} = \sum_{\substack{m \geq 1 \\ 4\nmid m}} \frac{mq^m}{1-q^m} = \sum_{\substack{m \geq 1 \\ 4\nmid m}} \sum_{n=1}^{\infty} mq^{mn}.$$

Hence,

$$\square(q)^4 = 1 + 8 \sum_{\substack{m \geq 1 \\ 4\nmid m}} \sum_{n=1}^{\infty} mq^{mn}. \tag{16.5}$$

Therefore, for any $N \geq 1$, the coefficient of q^N in $\square(q)^4$ is given by $8 \sum m$, where $mn = N$ for some positive integer n and 4 does not divide m. In other words, $\square_4(N)$ equals 8 times the sum of divisors of N not divisible by 4. $\square$

The result that any integer is a sum of four squares is the best possible, because not every integer is a sum of three squares, for example, 7. However, we can still say something about the number of ways an integer can be expressed as a sum of two squares, i.e., $\square_2(N)$.

Theorem 16.2. *For any positive integer we have*

$$\square_2(N) = 4 \times (\text{number of positive divisors of } N \text{ congruent to 1 modulo 4})$$
$$-4 \times (\text{number of positive divisors of } N \text{ congruent to 3 modulo 4}).$$
$$\tag{16.6}$$

Proof. This time, we let z tend to $i = \sqrt{-1}$ in (15.8). For the right-hand side, we have

$$\text{RHS} = \prod_{n=1}^{\infty} \frac{(1-q^n)^2(1+q^n)^2}{(1+iq^n)^2(1-iq^n)^2} = \left(\prod_{n=1}^{\infty} \frac{1-q^{2n}}{1+q^{2n}} \right)^2 = \square(-q^2)^2.$$

The left-hand side becomes

$$\text{LHS} = 1 - i \sum_{m,n=1}^{\infty} q^{mn}\left(i^{m+n} - (-i)^{m+n}\right).$$

Since $(-i)^{m+n} = i^{m+n}$ if $m+n$ is even, and $(-i)^{m+n} = -i^{m+n}$ if $m+n$ is odd, i.e., m and n have different parities, we have

$$\text{LHS} = 1 - 2i \sum_{\substack{m\geq 1 \\ m \text{ odd}}} \sum_{\substack{n\geq 1 \\ n \text{ even}}} q^{mn} i^{m+n} - 2i \sum_{\substack{m\geq 1 \\ m \text{ even}}} \sum_{\substack{n\geq 1 \\ n \text{ odd}}} q^{mn} i^{m+n}.$$

The two sums are identical, since each of them is symmetric in m and n. Replacing q^2 by $-q$ on both sides, we obtain

$$\begin{aligned}
\square(q)^2 &= 1 - 4 \sum_{\substack{m\geq 1 \\ m \text{ odd}}} \sum_{\substack{n\geq 1 \\ n \text{ even}}} (-q)^{mn/2} i^{m+n+1} \\
&= 1 - 4 \sum_{\substack{m\geq 1 \\ m \text{ odd}}} \sum_{\substack{n\geq 1 \\ n \text{ even}}} (-1)^{(m+1)/2} q^{mn/2}.
\end{aligned}$$

Letting $n = 2k$, we get the following result:

$$\square(q)^2 = 1 + 4 \sum_{\substack{m\geq 1 \\ m \text{ odd}}} \sum_{k=1}^{\infty} (-1)^{(m-1)/2} q^{mk}. \tag{16.7}$$

Let us examine the right-hand side of (16.7). For any $N \geq 1$, a $4q^N$ term arises for each odd m that divides N and is congruent to 1 modulo 4, so that $\frac{(m-1)}{2}$ is even. Similarly, a $(-4q^N)$ term arises for each odd m that divides N and is congruent to 3 modulo 4, so that $\frac{(m-1)}{2}$ is odd. Hence, the coefficient of q^N is given by the right-hand side of (16.6), as desired. $\square$

Corollary 16.1 (Fermat Theorem). *An odd prime p can be represented as a sum of two squares if and only if $p \equiv 1 \mod 4$, and the representation is essentially unique.*

Proof. A prime p has two divisors, 1 and p. If $p \equiv 1 \mod 4$, they are both congruent to 1 modulo 4. By Theorem 16.2, $\square_2(p) = 8$. Suppose $p = a^2 + b^2$. Each of the eight ordered pairs $(\pm a, \pm b)$, $(\pm b, \pm a)$ has p as the sum of squares. Also, these eight pairs are distinct, because p being odd implies $|a| \neq |b|$. Hence, they are all the possible cases, and the representation is unique up to the sign and the order. If $p \equiv -1 \mod 4$, Theorem 16.2 tells us that $\square_2(p) = 0$. $\square$

17

Explicit Formulas for Sums of Two and of Four Triangular Numbers

Besides partitions into square numbers, the Ramanujan product formula can also be applied to the study of partitions into sums of two or four triangular numbers. Let us recall the definition of the nth triangular number, introduced in Chapter 12:

$$\Delta_n = \frac{n(n+1)}{2}.$$

Since $\Delta_{-n-1} = \Delta_n$, the bilateral sequence $\{\Delta_n\}_{n\in\mathbb{Z}}$, is symmetric, and we shall restrict our definition of "triangular numbers" to $n \geq 0$ only. As in the case of square numbers, we define the following, similar, power series:

$$\Delta(q) = \sum_{n=0}^{\infty} q^{\Delta_n}.$$

(Unlike (16.1), the summation is taken over nonnegative integers only.) Then, the number of ways to express N as a sum of m triangular numbers, counting the order of summands, is equal to the coefficient of q^N in the power series $\Delta(q)^m$, and is denoted by $\Delta_m(N)$. The reason is similar to that for sums of square numbers.

Theorem 17.1. *For any positive integer N we have*

$$\Delta_2(N) = \textit{number of positive divisors of } 4N+1 \textit{ congruent to 1 modulo 4}$$
$$-\textit{number of positive divisors of } 4N+1 \textit{ congruent to 3 modulo 4}.$$

$$(17.1)$$

Proof. If we replace q by $-q$ and z by $-\sqrt{q}$ in (15.7), where $0 < q < 1$, we have, by (12.8),

$$
\begin{aligned}
\text{RHS} &= \prod_{n=1}^{\infty} \frac{\left(1-(-1)^n q^n\right)^2 \left(1-(-1)^n q^{n-1}\right)\left(1+(-1)^n q^n\right)}{\left(1-(-1)^n q^{n-\frac{1}{2}}\right)^2 \left(1+(-1)^n q^{n-\frac{1}{2}}\right)^2} \\
&= 2 \prod_{n=1}^{\infty} \frac{\left(1-(-1)^n q^n\right)^2 \left(1+(-1)^n q^n\right)^2}{\left(1-(-1)^n q^{n-\frac{1}{2}}\right)^2 \left(1+(-1)^n q^{n-\frac{1}{2}}\right)^2} \\
&= 2 \prod_{n=1}^{\infty} \frac{(1-q^{2n})^2}{(1-q^{2n-1})^2} = 2\Delta(q)^2,
\end{aligned}
$$

and

$$
\begin{aligned}
\text{LHS} &= \sum_{m,n=0}^{\infty} (-1)^{mn+m+n} q^{mn+\frac{m+n}{2}} - \sum_{m,n=1}^{\infty} (-1)^{mn-m-n} q^{mn-\frac{m+n}{2}} \\
&= \sum_{m,n=1}^{\infty} (-1)^{mn-1} q^{mn-\frac{m+n}{2}} - \sum_{m,n=1}^{\infty} (-1)^{mn-m-n} q^{mn-\frac{m+n}{2}},
\end{aligned}
$$

where we have replaced m by $m-1$ and n by $n-1$ in the first summation. When $m+n$ is odd, the corresponding terms in the two summations cancel. Hence, we have

$$
\begin{aligned}
\text{LHS} &= 2 \sum_{\substack{m,n\geq 1 \\ m+n \text{ even}}} (-1)^{mn-1} q^{mn-\frac{m+n}{2}} \\
&= 2 \sum_{\substack{m,n\geq 1 \\ m,n \text{ odd}}} q^{mn-\frac{m+n}{2}} - 2 \sum_{\substack{m,n\geq 1 \\ m,n \text{ even}}} q^{mn-\frac{m+n}{2}},
\end{aligned}
$$

and therefore,

$$
\Delta(q)^2 = \sum_{\substack{m,n\geq 1 \\ m,n \text{ odd}}} q^{mn-\frac{m+n}{2}} - \sum_{\substack{m,n\geq 1 \\ m,n \text{ even}}} q^{mn-\frac{m+n}{2}}. \tag{17.2}
$$

A $\pm q^N$ term appears on the right if and only if $N = mn - \frac{m+n}{2}$, or $4N + 1 = (2m-1)(2n-1)$, for some $m > 0$ and $n > 0$ either both odd or both even. If they are both odd, $2m - 1 \equiv 1 \mod 4$, and, if they are both even, $2m - 1 \equiv 3 \mod 4$. Hence, each factor of $4N + 1$ congruent to 1 modulo 4 contributes $+1$, and each factor of $4N + 1$ congruent to 3 modulo 4 contributes -1, to the coefficient of q^N. This completes the proof. $\square$

In particular, when $4N + 1$ is a prime number, we have the following corollary.

Corollary 17.1. *If N is a positive integer such that $4N+1$ is a prime number, then N can be represented uniquely as a sum of two distinct triangular numbers, up to reordering the summands.*

Proof. It is clear that if $4N + 1$ is prime, all the divisors of $4N + 1$ are 1 and $4N + 1$, both congruent to 1 modulo 4, and Theorem 17.1 implies $\Delta_2(N) = 2 - 0 = 2$. Note that by virtue of its definition as the coefficients of $\Delta(q)^m$, Δ_m counts any reordering of distinct summands. Hence, $\Delta_2(N) = 2$ implies that all the possible ways to represent N as a sum of two triangular numbers are either

$$N = \Delta_k + \Delta_l = \Delta_l + \Delta_k, \quad k \neq l,$$

or,

$$N = \Delta_k + \Delta_k = \Delta_l + \Delta_l, \quad k \neq l.$$

The second case is obviously invalid since the sequence $\{\Delta_n\}_{n \geq 0}$ is strictly increasing. The proof is thus complete. An alternative way to reject the second case is to note that $N = 2\Delta_k$ implies $4N + 1 = (2k + 1)^2$, which is not prime. $\square$

Examples of the corollary are $7 = 1 + 6$, $13 = 3 + 10$, and $43 = 15 + 28$. Another theorem concerns partitions into four triangular numbers.

Theorem 17.2. *For any positive integer N we have*

$$\Delta_4(N) = \text{sum of all divisors of } 2N + 1. \tag{17.3}$$

Proof. If we divide both sides of (15.7) by $1 - qz^{-2}$, replace q by q^2, and let z tend to q, we have

$$\text{RHS} = \prod_{n=1}^{\infty} \frac{(1 - q^{2n})^2 (1 - q^{2n})(1 - q^{2n})}{(1 - q^{2n-1})^2 (1 - q^{2n-1})^2} = \Delta(q)^4,$$

by (12.8). Since the right-hand side is finite, so is the left-hand side. Therefore, we may apply L'Hospital's rule:

$$
\begin{aligned}
\text{LHS} &= \lim_{z \to q} \frac{1}{1 - q^2 z^{-2}} \left(\sum_{m,n=0}^{\infty} q^{2mn} z^{m+n} - \sum_{m,n=1}^{\infty} q^{2mn} z^{-m-n} \right) \\
&= \frac{1}{2q^{-1}} \left(\sum_{m,n=0}^{\infty} (m+n) q^{2mn+m+n-1} + \sum_{m,n=1}^{\infty} (m+n) q^{2mn-m-n-1} \right) \\
&= \frac{1}{2} \sum_{m,n=1}^{\infty} (m+n-2) q^{2mn-m-n} + \sum_{m,n=1}^{\infty} (m+n) q^{2mn-m-n} \\
&= \sum_{m,n=1}^{\infty} (m+n-1) q^{2mn-m-n}.
\end{aligned}
$$

Since the expression is symmetric in m and n, we may rewrite the left-hand side and obtain

$$\Delta(q)^4 = \sum_{m,n=1}^{\infty} (2m - 1) q^{2mn-m-n}.$$

Letting $k = 2m - 1$ and $\ell = 2n - 1$, we get

$$\Delta(q)^4 = \sum_{\substack{k,\ell \leq 1 \\ k,\ell \text{ odd}}} kq^{\frac{(k\ell-1)}{2}}. \tag{17.4}$$

A q^N term appears in the sum if and only if $N = \frac{(k\ell-1)}{2}$, or $2N + 1 = k\ell$, for some odd numbers k and ℓ. Since every divisor of $2N + 1$ is odd, the coefficient of q^N is

$$\sum_{k|2N+1} k. \qquad \square$$

18
q-Antiderivative

After studying various applications, let us return to q-calculus. So far, we have talked about quantum differentiation only. What about quantum integration? Let us first consider the q-antiderivative.

Definition. The function $F(x)$ is a q-antiderivative of $f(x)$ if $D_q F(x) = f(x)$. It is denoted by

$$\int f(x) d_q x. \tag{18.1}$$

Note that we say "a" q-antiderivative instead of "the" q-antiderivative, because, as in ordinary calculus, an antiderivative is not unique. In ordinary calculus, the uniqueness is up to adding a constant, since the derivative of a function vanishes if and only if it is constant. The situation in quantum calculus is more subtle. $D_q \varphi(x) = 0$ if and only if $\varphi(qx) = \varphi(x)$, which does not necessarily imply φ a constant. Adding such a function φ does not alter the q-derivative of a function. However, if we require φ to be a formal power series, the condition $\varphi(qx) = \varphi(x)$ implies $q^n c_n = c_n$ for each n, where c_n is the coefficient of x^n. It is possible only when $c_n = 0$ for any $n \geq 1$, i.e., φ is constant. Therefore, if

$$f(x) = \sum_{n=0}^{\infty} a_n x^n$$

is a formal power series, then among formal power series, $f(x)$ has a unique q-antiderivative up to a constant term, which is

$$\int f(x)d_q x = \sum_{n=0}^{\infty} \frac{a_n x^{n+1}}{[n+1]} + C. \tag{18.2}$$

If $f(x)$ is a general function, we can still enhance the uniqueness by imposing some restrictions on the q-antiderivative. Consider again the function $\varphi(x)$, which has a zero q-derivative. The condition $\varphi(qx) = \varphi(x)$ is similar to that for a periodic function, but the period is smaller as x is closer to 0. To see this, suppose $q = 0.1$. Then, examples of periods are $(.1, 1], (.01, .1], (.001, .01]$, etc. If the graph of φ in $(.1, 1]$ is a straight but not horizontal line, in the periods closer to 0, the graph has the same shape, but it gets steeper and steeper, making φ discontinuous at $x = 0$. The general idea is contained in the next proposition.

Proposition 18.1. *Let $0 < q < 1$. Then, up to adding a constant, any function $f(x)$ has at most one q-antiderivative that is continuous at $x = 0$.*

Proof. Suppose F_1 and F_2 are two q-antiderivatives of f that are continuous at 0. Let $\varphi = F_1 - F_2$. The function φ is also continuous at 0, and has the property $\varphi(qx) = \varphi(x)$ for any x, since $D_q \varphi = 0$. For some $A > 0$, let

$$
\begin{aligned}
m &= \inf\{\varphi(x) | qA \le x \le A\}, \\
M &= \sup\{\varphi(x) | qA \le x \le A\},
\end{aligned}
$$

which may be infinity if φ is unbounded above and/or below.

Assuming $m < M$, at least one of $\varphi(0) \neq m$ and $\varphi(0) \neq M$ is true. Suppose $\varphi(0) \neq m$. By continuity at $x = 0$, given $\epsilon > 0$ small enough, we can always find a $\delta > 0$ such that

$$m + \epsilon \notin \varphi(0, \delta).$$

On the other hand, $q^N A < \delta$ for some sufficiently large N. Since $\varphi(qx) = \varphi(x)$, we have

$$m + \epsilon \in (m, M) \subset \varphi[qA, A] = \varphi[q^{N+1}A, q^N A] \subset \varphi(0, \delta),$$

leading to a contradiction. Therefore, $m = M$ and φ is constant in $[qA, A]$, which means it is constant everywhere. $\square$

The proposition tells us that the uniqueness of the q-antiderivative is substantially improved by requiring continuity at $x = 0$. The existence problem will be discussed in the next chapter.

We conclude this chapter with the following formula for the change of variable $u = u(x) = \alpha x^{\beta}$, where α and β are constants. Suppose that $F(x)$ is a q-antiderivative of $f(x)$. Then

$$\int f(u)d_q u = F(u) = F(u(x)).$$

We have for any q', using (1.15),

$$
\begin{aligned}
F(u(x)) &= \int D_{q'} F(u(x)) d_{q'} x \\
&= \int (D_{q'^\beta} F)(u(x)) \cdot D_{q'} u(x) \, d_{q'} x \\
&= \int (D_{q'^\beta} F)(u(x)) \, d_{q'} u(x).
\end{aligned}
$$

Choosing $q' = q^{1/\beta}$, we have $D_{q'^\beta} F = D_q F = f$, and thus

$$
\int f(u) d_q u = \int f(u(x)) d_{q^{1/\beta}} u(x). \tag{18.3}
$$

This formula means that $f(u(x)) D_{q^{1/\beta}} u(x)$ is one of the q-antiderivatives of $f(u)$.

19
Jackson Integral

Suppose $f(x)$ is an arbitrary function. To construct its q-antiderivative $F(x)$, recall the operator $\hat{M}_q$, defined by $\hat{M}_q(F(x)) = F(qx)$ in Chapter 5. Then we have by the definition of a q-derivative:

$$\frac{1}{(q-1)x}(\hat{M}_q - 1)F(x) = \frac{F(qx) - F(x)}{(q-1)x} = f(x). \qquad (19.1)$$

Note that the order is important, because operators do not commute. We can then *formally* write the q-antiderivative as

$$F(x) = \frac{1}{1 - \hat{M}_q}\left((1-q)xf(x)\right) = (1-q)\sum_{j=0}^{\infty}\hat{M}_q^j\left(xf(x)\right),$$

using the geometric series expansion, and thus we get

$$\int f(x)d_qx = (1-q)x\sum_{j=0}^{\infty}q^j f(q^j x). \qquad (19.2)$$

This series is called the *Jackson integral* of $f(x)$. From this definition one easily derives a more general formula:

$$\begin{aligned}
\int f(x)D_q g(x)d_q x &= (1-q)x\sum_{j=0}^{\infty}q^j f(q^j x)D_q g(q^j x) \\
&= (1-q)x\sum_{j=0}^{\infty}q^j f(q^j x)\frac{g(q^j x) - g(q^{j+1}x)}{(1-q)q^j x},
\end{aligned}$$

or

$$\int f(x)d_q g(x) = \sum_{j=0}^{\infty} f(q^j x)\Big(g(q^j x) - g(q^{j+1} x)\Big). \qquad (19.3)$$

We have merely derived (19.2) formally, and have yet to examine under what conditions it really converges to a q-antiderivative. The theorem below gives a sufficient condition for this.

Theorem 19.1. *Suppose $0 < q < 1$. If $|f(x)x^\alpha|$ is bounded on the interval $(0, A]$ for some $0 \le \alpha < 1$, then the Jackson integral defined by (19.2) converges to a function $F(x)$ on $(0, A]$, which is a q-antiderivative of $f(x)$. Moreover, $F(x)$ is continuous at $x = 0$ with $F(0) = 0$.*

Proof. Suppose $|f(x)x^\alpha| < M$ on $(0, A]$. For any $0 < x \le A$, $j \ge 0$,

$$|f(q^j x)| < M(q^j x)^{-\alpha}.$$

Thus, for any $0 < x \le A$, we have

$$|q^j f(q^j x)| < Mq^j (q^j x)^{-\alpha} = Mx^{-\alpha}(q^{1-\alpha})^j. \qquad (19.4)$$

Since $1 - \alpha > 0$ and $0 < q < 1$, we see that our series is majorized by a convergent geometric series. Hence, the right-hand side of (19.2) converges pointwise to some function $F(x)$. It follows directly from (19.2) that $F(0) = 0$. The fact that $F(x)$ is continuous at $x = 0$, i.e., $F(x)$ tends to zero as $x \to 0$, is clear if we consider, using (19.4),

$$\left| (1 - q)x \sum_{j=0}^{\infty} q^j f(q^j x) \right| < \frac{M(1 - q)x^{1-\alpha}}{1 - q^{1-\alpha}}, \qquad 0 < x \le A.$$

To verify that $F(x)$ is a q-antiderivative, we q-differentiate it:

$$
\begin{aligned}
D_q F(x) &= \frac{1}{(1-q)x}\left((1-q)x \sum_{j=0}^{\infty} q^j f(q^j x) - (1-q)qx \sum_{j=0}^{\infty} q^j f(q^{j+1} x) \right) \\
&= \sum_{j=0}^{\infty} q^j f(q^j x) - \sum_{j=0}^{\infty} q^{j+1} f(q^{j+1} x) \\
&= \sum_{j=0}^{\infty} q^j f(q^j x) - \sum_{j=1}^{\infty} q^j f(q^j x) = f(x).
\end{aligned}
$$

Note that if $x \in (0, A]$ and $0 < q < 1$, then $qx \in (0, A]$, and the q-differentiation is valid. $\square$

By Proposition 18.1, if the assumption of Theorem 19.1 is satisfied, the Jackson integral gives the unique q-antiderivative that is continuous at $x = 0$, up to adding a constant. On the other hand, if we know that $F(x)$ is a q-antiderivative of $f(x)$ and $F(x)$ is continuous at $x = 0$, $F(x)$ must be given, up to adding a constant, by Jackson's formula (19.2), since a partial sum of the Jackson integral is

$$
\begin{aligned}
(1-q)x \sum_{j=0}^{N} q^j f(q^j x) &= (1-q)x \sum_{j=0}^{N} q^j \left. D_q F(t)\right|_{t=q^j x} \\
&= (1-q)x \sum_{j=0}^{N} q^j \left(\frac{F(q^j x) - F(q^{j+1}x)}{(1-q)q^j x} \right) \\
&= \sum_{j=0}^{N} \left(F(q^j x) - F(q^{j+1}x) \right) = F(x) - F(q^{N+1}x),
\end{aligned}
$$

which tends to $F(x) - F(0)$ as $N \to \infty$, by the continuity of $F(x)$ at $x = 0$.

To see an example where the Jackson formula fails, consider $f(x) = 1/x$. Since

$$
D_q \log x = \frac{\log(qx) - \log(x)}{(q-1)x} = \frac{\log q}{q-1} \frac{1}{x}, \tag{19.5}
$$

we have

$$
\int \frac{1}{x} d_q x = \frac{q-1}{\log q} \log x. \tag{19.6}
$$

However, the Jackson formula gives

$$
\int \frac{1}{x} d_q x = (1-q) \sum_{j=0}^{\infty} 1 = \infty.
$$

The formula fails because $f(x)x^\alpha$ is not bounded for any $0 \le \alpha < 1$. Note that $\log x$ is not continuous at $x = 0$.

Now we apply the Jackson formula (19.2) to define the definite q-integral.

Definition. Suppose $0 < a < b$. The definite q-integral is defined as

$$
\int_0^b f(x)d_q x = (1-q)b \sum_{j=0}^{\infty} q^j f(q^j b) \tag{19.7}
$$

and

$$
\int_a^b f(x)d_q x = \int_0^b f(x)d_q x - \int_0^a f(x)d_q x. \tag{19.8}
$$

As before (see 19.3), we derive from (19.7) a more general formula:

$$
\int_0^b f(x)d_q g(x) = \sum_{j=0}^{\infty} f(q^j b)\left(g(q^j b) - g(q^{j+1}b) \right). \tag{19.9}
$$

Note that this definition conforms to the fact that the Jackson integral vanishes at $x = 0$. Geometrically, the integral in (19.7) corresponds to the area of the union of an infinite number of rectangles, as drawn below.

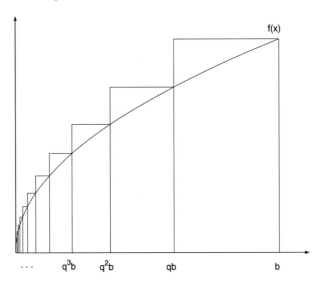

On $[\epsilon, b]$, where ϵ is a small positive number, the sum consists of finitely many terms, and is in fact a Riemann sum. Therefore, as $q \to 1$, the width of the rectangles approaches zero, and the sum tends to the Riemann integral on $[\epsilon, b]$. Since ϵ is arbitrary, we thus have, provided that $f(x)$ is continuous in the interval $[0, b]$,

$$\lim_{q \to 1} \int_0^b f(x) d_q x = \int_0^b f(x) dx. \tag{19.10}$$

We cannot obtain a good definition of improper integral by simply letting $b \to \infty$ in (19.7). Instead, since

$$
\int_{q^{j+1}}^{q^j} f(x) d_q x = \int_0^{q^j} f(x) d_q x - \int_0^{q^{j+1}} f(x) d_q x
$$

$$
= (1-q) \sum_{k=0}^{\infty} q^{j+k} f(q^{j+k}) - (1-q) \sum_{k=0}^{\infty} q^{j+k+1} f(q^{j+k+1}),
$$

and thus

$$\int_{q^{j+1}}^{q^j} f(x) d_q x = (1-q) q^j f(q^j), \tag{19.11}$$

it is natural to define the improper q-integral as follows.

Definition. The improper q-integral of $f(x)$ on $[0, +\infty)$ is defined to be

$$\int_0^\infty f(x) d_q x = \sum_{j=-\infty}^{\infty} \int_{q^{j+1}}^{q^j} f(x) d_q x \tag{19.12}$$

if $0 < q < 1$, or

$$\int_0^\infty f(x)d_q x = \sum_{j=-\infty}^\infty \int_{q^j}^{q^{j+1}} f(x)d_q x$$

if $q > 1$.

Proposition 19.1. *The improper q-integral defined above converges if $x^\alpha f(x)$ is bounded in a neighborhood of $x = 0$ with some $\alpha < 1$ and for sufficiently large x with some $\alpha > 1$.*

Proof. By (19.11), we have

$$\int_0^\infty f(x)d_q x = |1 - q| \sum_{j=-\infty}^\infty q^j f(q^j). \tag{19.13}$$

Since the summation

$$\sum_{j=-\infty}^\infty q^j f(q^j) = \sum_{j=0}^\infty q^j f(q^j) + \sum_{j=1}^\infty q^{-j} f(q^{-j})$$

remains unchanged if we replace q by q^{-1}, it suffices to consider the case of $q < 1$. The convergence of the first sum is proved by Theorem 19.1. For the second sum, suppose for large x we have $|x^\alpha f(x)| < M$ where $\alpha > 1$ and $M > 0$. Then, we have for sufficiently large j,

$$|q^{-j} f(q^{-j})| = q^{j(\alpha-1)} |q^{-j\alpha} f(q^{-j})| < M q^{j(\alpha-1)}.$$

Therefore the second sum is also majorized by a convergent geometric series, and thus converges.   □

Now let us discuss the change of variables $u = u(x) = \alpha x^\beta$ in definite integrals. If the Jackson integral of a function converges, the Jackson formula may be used to rederive (18.3). Indeed, consider its right-hand side:

$$\begin{aligned}
\text{RHS} &= \sum_{j=0}^\infty f(u(q^{j/\beta}x))\left(u(q^{j/\beta}x) - u(q^{(j+1)/\beta})\right) \\
&= \sum_{j=0}^\infty f(\alpha q^j x^\beta)\left(\alpha q^j x^\beta - \alpha q^{j+1} x^\beta\right) \\
&= \sum_{j=0}^\infty f(q^j u)(q^j u - q^{j+1} u) = (1 - q)u \sum_{j=0}^\infty q^j f(q^j u) = \text{LHS},
\end{aligned}$$

where we have used (19.3). Replacing x by a and b above, one readily obtains that

$$\int_{u(a)}^{u(b)} f(u)d_q u = \int_a^b f(u(x))d_{q^{1/\beta}} u(x). \tag{19.14}$$

With the Newton–Leibniz formula (20.1) to be introduced in the next chapter, (19.14) can be shown even more directly, since both of its sides equal $F(u(b)) - F(u(a))$, where F is a q-antiderivative of f continuous at $x = 0$. Since (20.1) is true for improper integrals, we see that if $\alpha, \beta > 0$, so that $u(+\infty) = +\infty$, (19.14) is also true for $b = +\infty$. In particular, we have for $\alpha > 0, \beta = 1$,

$$\int_0^\infty f(\alpha x)d_q x = \frac{1}{\alpha}\int_0^\infty f(x)d_q x. \tag{19.15}$$

20
Fundamental Theorem of q-Calculus and Integration by Parts

In ordinary calculus, a derivative is defined as the limit of a ratio, and a definite integral is defined as the limit of an infinite sum. Their subtle and surprising relation is given by the Newton–Leibniz formula, also called the fundamental theorem of calculus. In contrast, since the introduction of the definite q-integral has been motivated by an antiderivative, the relation between the q-derivative and definite q-integral is more obvious. Analogous to the ordinary case, we have the following *fundamental theorem*, or *Newton–Leibniz formula*, for q-calculus.

Theorem 20.1. *(Fundamental theorem of q-calculus) If $F(x)$ is an antiderivative of $f(x)$ and $F(x)$ is continuous at $x = 0$, we have*

$$\int_a^b f(x)d_qx = F(b) - F(a), \tag{20.1}$$

where $0 \leq a < b \leq \infty$.

Proof. As noted in the previous chapter, since $F(x)$ is continuous at $x = 0$, $F(x)$ is given by the Jackson formula, up to adding a constant, i.e.,

$$F(x) = (1 - q)x \sum_{j=0}^{\infty} q^j f(q^j x) + F(0).$$

Since by definition,

$$\int_0^a f(x)d_qx = (1 - q)a \sum_{j=0}^{\infty} q^j f(q^j a),$$

we have

$$\int_0^a f(x)d_qx = F(a) - F(0).$$

Similarly, we have, for finite b,

$$\int_0^b f(x)d_qx = F(b) - F(0),$$

and thus

$$\int_a^b f(x)d_qx = \int_0^b f(x)d_qx - \int_0^a f(x)d_qx = F(b) - F(a).$$

Putting $a = q^{j+1}$ (or q^j) and $b = q^j$ (or q^{j+1}), where $0 < q < 1$ (or $q > 1$), and considering the definition of improper q-integral (19.12), we see that (20.1) is true for $b = \infty$ as well if $\lim_{x\to\infty} F(x)$ exists.   □

Corollary 20.1. *If $f'(x)$ exists in a neighborhood of $x = 0$ and is continuous at $x = 0$, where $f'(x)$ denotes the ordinary derivative of $f(x)$, we have*

$$\int_a^b D_qf(x)d_qx = f(b) - f(a). \tag{20.2}$$

Proof. Using L'Hospital's rule, we get

$$\lim_{x\to 0} D_qf(x) = \lim_{x\to 0} \frac{f(qx) - f(x)}{(q-1)x} = \lim_{x\to 0} \frac{qf'(qx) - f'(x)}{q-1} = f'(0).$$

Hence $D_qf(x)$ can be made continuous at $x = 0$ if we define $(D_qf)(0) = f'(0)$, and (20.2) follows from the theorem.   □

An important difference between the definite q-integral and its ordinary counterpart is that even if we are integrating a function on an interval like $[1, 2]$, we have to care about its behavior at $x = 0$. This has to do with the definition of the definite q-integral and the condition for the convergence of the Jackson integral.

Now suppose $f(x)$ and $g(x)$ are two functions whose ordinary derivatives exist in a neighborhood of $x = 0$ and are continuous at $x = 0$. Using the product rule (1.12), we have

$$D_q\Big(f(x)g(x)\Big) = f(x)\Big(D_qg(x)\Big) + g(qx)\Big(D_qf(x)\Big).$$

Since the product of differentiable functions is also differentiable in ordinary calculus, we can apply Corollary 20.1 to obtain

$$f(b)g(b) - f(a)g(a) = \int_a^b f(x)\Big(D_qg(x)\Big)d_qx + \int_a^b g(qx)\Big(D_qf(x)\Big)d_qx,$$

or

$$\int_a^b f(x)d_qg(x) = f(b)g(b) - f(a)g(a) - \int_a^b g(qx)d_qf(x), \tag{20.3}$$

which is the formula of q-integration by parts. Note that $b = \infty$ is allowed as well.

The q-integration by parts can be applied to obtain the q-Taylor formula with the Cauchy remainder term.

Theorem 20.2. *Suppose $D_q{}^j f(x)$ is continuous at $x = 0$ for any $j \leq n+1$. Then, we have a q-analogue of Taylor's formula with the Cauchy remainder:*

$$f(b) = \sum_{j=0}^{n} \left(D_q{}^j f \right)(a) \frac{(b-a)_q^j}{[j]!} + \frac{1}{[n]!} \int_a^b D_q{}^{n+1} f(x)(b-qx)_q^n d_q x. \quad (20.4)$$

Proof. Since $D_q f(x)$ is continuous at $x = 0$, by Theorem 20.1 we have

$$f(b) - f(a) = \int_a^b D_q f(x) d_q x = -\int_a^b D_q f(x) d_q (b-x),$$

which proves (20.4) in the case where $n = 0$. Assume that (20.4) holds for $n - 1$:

$$f(b) = \sum_{j=0}^{n-1} \left(D_q{}^j f \right)(a) \frac{(b-a)_q^j}{[j]!} + \frac{1}{[n-1]!} \int_a^b D_q{}^n f(x)(b-qx)_q^{n-1} d_q x.$$

Using (3.11) and applying q-integration by parts (20.3), we obtain

$$\begin{aligned}
\int_a^b D_q{}^n f(x)(b-qx)_q^{n-1} d_q x &= -\frac{1}{[n]} \int_a^b D_q{}^n f(x) d_q (b-x)_q^n \\
&= D_q{}^n f(a) \frac{(b-a)_q^n}{[n]} \\
&\quad + \frac{1}{[n]} \int_a^b (b-qx)_q^n D_q{}^{n+1} f(x) d_q x,
\end{aligned}$$

and the proof is complete by induction. $\square$

21

q-Gamma and q-Beta Functions

Being related to solutions of special types of differential equations, many important functions in analysis are defined in terms of definite integrals. The following two functions, introduced by Euler,

$$\Gamma(t) \quad = \quad \int_0^\infty x^{t-1} e^{-x} dx, \quad t > 0, \tag{21.1}$$

$$B(t, s) \quad = \quad \int_0^1 x^{t-1}(1-x)^{s-1} dx, \quad s, t > 0, \tag{21.2}$$

and called the *gamma* and the *beta functions* respectively, are the most important examples. Some of their properties are listed below:

$$\Gamma(t+1) \quad = \quad t\Gamma(t), \tag{21.3}$$

$$\Gamma(n) \quad = \quad (n-1)! \quad \text{if } n \text{ is a positive integer}, \tag{21.4}$$

$$B(t, s) \quad = \quad \frac{\Gamma(t)\Gamma(s)}{\Gamma(t+s)}. \tag{21.5}$$

In particular, (21.4) tells us that the gamma function may be regarded as a generalization of factorials. In this chapter, we study the q-analogues of these two functions and their various properties, including the q-analogues of (21.3)–(21.5). We shall assume that $0 < q < 1$.

Definition. For any $t > 0$,

$$\Gamma_q(t) = \int_0^\infty x^{t-1} E_q^{-qx} d_q x \tag{21.6}$$

is called the *q-gamma function*.

First we note that by (9.10), $E_q^0 = 1$, and by (9.7) and (9.13), $E_q^{-\infty} = \lim_{x\to\infty} 1/e_q^x = 0$. Using (9.11) and q-integration by parts (20.3), we have

$$\int_0^\infty x^t E_q^{-qx} d_q x = -\int_0^\infty x^t d_q E_q^{-x} = [t] \int_0^\infty x^{t-1} E_q^{-qx} d_q x,$$

and hence,

$$\Gamma_q(t+1) = [t]\Gamma_q(t), \tag{21.7}$$

for any $t > 0$. Since

$$\Gamma_q(1) = \int_0^\infty E_q^{-qx} d_q x = E_q^0 - E_q^{-\infty} = 1,$$

we have for any nonnegative integer n,

$$\Gamma_q(n+1) = [n]!. \tag{21.8}$$

To study the gamma function at $t \notin \mathbb{N}$, it is helpful to consider a seemingly more complicated function.

Definition. For any $t, s > 0$,

$$B_q(t, s) = \int_0^1 x^{t-1}(1 - qx)_q^{s-1} d_q x \tag{21.9}$$

is called the q-beta function.

By the definitions of proper and improper integrals, (19.7) and (19.12), we have

$$
\begin{aligned}
B_q(t, \infty) &= (1-q)\sum_{j=0}^{\infty} q^j (q^j a)^{t-1}(1 - q^{j+1})_q^\infty \\
&= (1-q)\sum_{j=-\infty}^{\infty} q^j (q^j a)^{t-1}(1 - q^{j+1})_q^\infty \\
&= \int_0^\infty x^{t-1}(1 - qx)_q^\infty d_q x,
\end{aligned}
$$

where we have used the fact that $(1 - q^{j+1})_q^\infty = 0$ for any negative integer j. The relation between $\Gamma_q(t)$ and $B_q(t, s)$ is revealed by the formula $E_q^x = (1 + (1-q)x)_q^\infty$. Thus, we have

$$B_q(t, \infty) = \int_0^\infty x^{t-1} E_q^{-\frac{qx}{(1-q)}} d_q x,$$

and performing the change of variable $x = (1 - q)y$ (19.15), we obtain

$$B_q(t, \infty) = (1-q)^t \int_0^\infty y^{t-1} E_q^{-qy} d_q y,$$

or

$$\Gamma_q(t) = \frac{B_q(t, \infty)}{(1 - q)^t}. \tag{21.10}$$

Introducing another variable s may seem to be a backward movement at first glance. However, it actually increases our freedom to manipulate the functions and simplify the problem.

Proposition 21.1. *(a) If $t > 0$ and n is a positive integer, we have*

$$B_q(t, n) = \frac{(1 - q)(1 - q)_q^{n-1}}{(1 - q^t)_q^n}. \tag{21.11}$$

(b) For any $t, s > 0$, we have

$$B_q(t, s) = \frac{(1 - q)(1 - q)_q^\infty (1 - q^{t+s})_q^\infty}{(1 - q^t)_q^\infty (1 - q^s)_q^\infty}. \tag{21.12}$$

Proof. (a) We first derive two recurrence relations for $B_q(t, s)$. Firstly, using (3.11) and q-integration by parts, we have, for any $t > 1$, $s > 0$,

$$B_q(t, s) = -\frac{1}{[s]} \int_0^1 x^{t-1} d_q (1 - x)_q^s = \frac{[t-1]}{[s]} \int_0^1 x^{t-2}(1 - qx)_q^s d_q x,$$

and hence

$$B_q(t, s) = \frac{[t-1]}{[s]} B_q(t - 1, s + 1). \tag{21.13}$$

On the other hand, we have

$$\begin{aligned}
B_q(t, n + 1) &= \int_0^1 x^{t-1}(1 - qx)_q^{n-1}(1 - q^n x) d_q x \\
&= \int_0^1 x^{t-1}(1 - qx)_q^{n-1} d_q x - q^n \int_0^1 x^t (1 - qx)_q^{n-1} d_q x,
\end{aligned}$$

and thus

$$B_q(t, n + 1) = B_q(t, n) - q^n B_q(t + 1, n). \tag{21.14}$$

Combining (21.13) and (21.14), we obtain

$$B_q(t, n + 1) = B_q(t, n) - \frac{q^n [t]}{[n]} B_q(t, n + 1),$$

or

$$B_q(t, n + 1) = \frac{1 - q^n}{1 - q^{t+n}} B_q(t, n), \tag{21.15}$$

for any $t > 0$ and positive integer n. Since

$$B_q(t, 1) = \int_0^1 x^{t-1} d_q x = \frac{1}{[t]},$$

we have

$$B_q(t, n) = \frac{(1 - q^{n-1}) \cdots (1 - q)}{(1 - q^{t+n-1}) \cdots (1 - q^{t+1})[t]} = \frac{(1 - q)(1 - q)_q^{n-1}}{(1 - q^t)_q^n},$$

as desired.

(b) We will use an argument similar to the one used in the proof of Theorem 13.1. By part (a), since

$$(1-q)_q^{n-1} = \frac{(1-q)_q^\infty}{(1-q^n)_q^\infty} \quad \text{and} \quad \frac{1}{(1-q^t)_q^n} = \frac{(1-q^{t+n})_q^\infty}{(1-q^t)_q^\infty},$$

(21.12) is true for $s = 1, 2, 3, \dots$. We may write the left-hand side of (21.12) as

$$\int_0^1 \frac{x^{t-1}(1-qx)_q^\infty}{(1-ax)_q^\infty} d_q x$$

and its right-hand side as

$$\frac{(1-q)(1-q)_q^\infty(1-aq^t)_q^\infty}{(1-q^t)_q^\infty(1-a)_q^\infty},$$

where $a = q^s$. Then, both sides are formal power series in q. Their corresponding coefficients are equal for infinitely many values of a, namely, $a = q, q^2, q^3, \dots$. However, since all coefficients are polynomials in a and distinct polynomials may coincide at only finitely many points, the two series have identical coefficients, and equality is established.   □

Part (a) of the proposition gives us an explicit expression for the q-gamma function. Using (21.10) and letting $n \to \infty$ in (21.11), we obtain

$$\Gamma_q(t) = \frac{(1-q)_q^\infty}{(1-q)^{t-1}(1-q^t)_q^\infty}. \tag{21.16}$$

Part (b) shows that the q-beta function is symmetric about t and s, namely, $B(t, s) = B(s, t)$, which is not obvious from simply looking at (21.9). Comparing (21.12) and (21.16), we also obtain an expression of the q-beta function in terms of the q-gamma function similar to (21.5):

$$B_q(t, s) = \frac{\Gamma_q(t)\Gamma_q(s)}{\Gamma_q(t+s)}. \tag{21.17}$$

This concludes our discussion of the q-gamma and the q-beta functions.

22

h-Derivative and h-Integral

We have thus far studied only q-calculus. Now we turn to h-calculus. Firstly, let us recall from Chapter 1 another quantum derivative that is characterized by an additive parameter h, the h-derivative:

$$D_h f(x) = \frac{f(x+h) - f(x)}{h},$$

where $h \neq 0$. Let us begin by developing the properties of h-calculus in an analogous way to what we have done for q-calculus, and discuss its applications in subsequent chapters.

It is easy to verify the product and quotient rules for h-differentiation:

$$D_h\Big(f(x)g(x)\Big) = f(x)D_h g(x) + g(x+h)D_h f(x), \qquad (22.1)$$

$$D_h\left(\frac{f(x)}{g(x)}\right) = \frac{g(x)D_h f(x) - f(x)D_h g(x)}{g(x)g(x+h)}. \qquad (22.2)$$

The first formula follows from (1.4), and the second one follows easily from the first. The resemblance in the product rule suggests that the h-binomial may be defined in a similar way.

Definition. The h-analogue of a binomial $(x-a)^n$ is

$$(x-a)_h^n = (x-a)(x-a-h)\cdots\big(x-a-(n-1)h\big) \qquad (22.3)$$

when $n \geq 1$, and $(x-a)_h^0 = 1$.

To verify that this is an adequate definition of h-binomial, consider

$$D_h(x-a)_h^n = \frac{1}{h}\Big((x-a+h)(x-a)\cdots(x-a-(n-2)h)$$
$$-(x-a)(x-a-h)\cdots(x-a-(n-1)h)\Big)$$
$$=(x-a)\cdots(x-a-(n-2)h)\frac{(x-a+h)-(x-a-(n-1)h)}{h}.$$

Hence, we have

$$D_h(x-a)_h^n = n(x-a)_h^{n-1}. \tag{22.4}$$

Note that the h-analogue of an integer n is still n, and $(x-0)_h^n \neq x^n$. With (22.4) above, the sequence of polynomials $\{(x-a)_h^n\}$ satisfies the conditions of Theorem 2.1 with respect to the linear operator $D \equiv D_h$. Therefore, we have the following h-Taylor formula for a polynomial $f(x)$ of degree N:

$$f(x) = \sum_{j=0}^{N} (D_h^j f)(a)\frac{(x-a)_h^j}{j!}. \tag{22.5}$$

Example. The h-Taylor formula applied to $f(x) = (x+b)^N$, $a=0$, gives:

$$(x+b)^N = \sum_{j=0}^{N} \binom{N}{j} b_h^{N-j} x_h^j.$$

The following facts are stated without proof. The proofs are similar to those already given for their q-versions:

$$(x-a)_h^{m+n} = (x-a)_h^n(x-a-nh)_h^m, \tag{22.6}$$
$$D_h(a-x)_h^n = -n(a-h-x)_h^{n-1}, \tag{22.7}$$
$$D_h\frac{1}{(x-a)_h^n} = -\frac{n}{(x+h-a)_h^{n+1}}, \tag{22.8}$$
$$D_h\frac{1}{(a-x)_h^n} = \frac{n}{(a-x)_h^{n+1}}. \tag{22.9}$$

The formulas above may be extended to all integers if we define

$$(x-a)_h^{-n} = \frac{1}{(x-a+nh)_h^n}, \tag{22.10}$$

as dictated by (22.6).

Next, let us discuss $f(x) = e_h^x$, the h-analogue of the exponential function. Three properties that $f(x)$ should have are (i) $f(0) = 1$, (ii) $D_h f(x) = f(x)$ for any x, and (iii) $f(x)$ admits h-Taylor expansion (22.5) about $x = 0$ (with $N = \infty$) for small h. In fact, these three properties uniquely characterize $f(x)$, since with (i) and (ii), we know that $(D_h^j f)(0) = 1$ for any

$j = 0, 1, \ldots$. By (22.5) with $a = 0$, we have

$$
\begin{aligned}
f(x) &= \sum_{j=0}^{\infty} \frac{(x-0)_h^j}{j!} = \sum_{j=0}^{\infty} \frac{x(x-h)\cdots(x-(j-1)h)}{j!} \\
&= \sum_{j=0}^{\infty} \frac{\frac{x}{h}(\frac{x}{h}-1)\cdots(\frac{x}{h}-j+1))h^j}{j!} = \sum_{j=0}^{\infty} \binom{x/h}{j} h^j,
\end{aligned}
$$

and hence, by ordinary Taylor expansion of a binomial, we have

$$
e_h^x = (1+h)^{\frac{x}{h}}. \tag{22.11}
$$

In particular, $e_1^x = 2^x$. Also, as $h \to 0$, the base $(1+h)^{\frac{1}{h}}$ approaches e, as expected. Note that

$$
D_h e_h^{\alpha x} = \frac{(1+h)^{\frac{\alpha(x+h)}{h}} - (1+h)^{\frac{\alpha x}{h}}}{h} = \frac{(1+h)^\alpha - 1}{h}(1+h)^{\frac{\alpha x}{h}},
$$

and hence

$$
D_h e_h^{\alpha x} = [\alpha]_{1+h} e_h^{\alpha x}, \tag{22.12}
$$

which serves as an example where the chain rule fails.

If $D_h F(x) = f(x)$, then $F(x)$ is called an h-antiderivative of $f(x)$ and is denoted by

$$
\int f(x) d_h x.
$$

The definite h-integral of a function from $x = a$ to $x = b$, where a and b differ by an integer multiple of h, may be defined as a finite sum:

Definition. If $b - a \in h\mathbb{Z}$, we define the *definite h-integral* to be

$$
\int_a^b f(x) d_h x = \begin{cases} h\Big(f(a) + f(a+h) + \cdots + f(b-h)\Big) & \text{if } a < b, \\ 0 & \text{if } a = b, \\ -h\Big(f(b) + f(b+h) + \cdots + f(a-h)\Big) & \text{if } a > b. \end{cases} \tag{22.13}
$$

With this definition, the definite h-integral is a Riemann sum of $f(x)$ on the interval $[a, b]$, which is partitioned into subintervals of equal width. The following theorem justifies (22.13) as an appropriate definition for the h-integral.

Theorem 22.1. *(Fundamental theorem of h-calculus) If $F(x)$ is an h-antiderivative of $f(x)$ and $b - a \in h\mathbb{Z}$, we have*

$$
\int_a^b f(x) d_h x = F(b) - F(a). \tag{22.14}
$$

Proof. If $b > a$, then by definition we have

$$
\int_a^b f(x)d_h x \;=\; h \sum_{j=0}^{\frac{(b-a)}{h}-1} f(a+jh) = h \sum_{j=0}^{\frac{(b-a)}{h}-1} D_h F(x)\big|_{x=a+jh}
$$

$$
=\; \sum_{j=0}^{\frac{(b-a)}{h}-1} \Big(F(a+(j+1)h) - F(a+jh) \Big) = F(b) - F(a),
$$

as desired. The case $b < a$ is similar, and the last case, $b = a$, is trivial.
□

Applying Theorem 22.1 to $D_h(F(x)g(x))$ and using (22.1), we obtain the h-version of integration by parts:

$$
\int_a^b f(x)d_h g(x) = f(b)g(b) - f(a)g(a) - \int_a^b g(x+h)d_h f(x), \quad (22.15)
$$

where (assuming $a < b$)

$$
\int_a^b f(x)d_h g(x) \;=\; \int_a^b f(x)D_h g(x)d_h x
$$

$$
=\; h \sum_{j=0}^{\frac{(b-a)}{h}-1} f(a+jh)\,(D_h g)\,(a+jh)
$$

$$
=\; \sum_{j=0}^{\frac{(b-a)}{h}-1} f(a+jh)\Big(g(a+jh+h) - g(a+jh) \Big).
$$

Take $h = 1$ and a, b to be integers, with $a < b$. For a function $\varphi(x)$ define

$$
f(x) = \varphi(0) + \varphi(1) + \cdots + \varphi(x-1),
$$

where x is a positive integer. In other words, $D_1 f(x) = \varphi(x)$. From (22.15), we obtain

$$
\sum_{j=a}^{b-1} \varphi(j)g(j+1) = g(b)f(b) - g(a)f(a) - \sum_{j=a}^{b-1} f(j)\,(g(j+1) - g(j)). \,(22.16)
$$

This formula is known as the *Abel transform*.

Another useful formula can be obtained by repeatedly applying h-integration by parts. If $x - a \in h\mathbb{Z}$, then using (22.7) and (22.15), we have

$$
\begin{aligned}
f(x) - f(a) &= \int_a^x D_h f(t)d_h t = -\int_a^x D_h f(t)d_h(x-t) \\
&= (D_h f)(a)(x-a) + \int_a^x (x-h-t)D_h^2 f(t)d_h t \\
&= (D_h f)(a)(x-a) - \frac{1}{2}\int_a^x D_h^2 f(t)d_h(x-t)_h^2 \\
&= (D_h f)(a)(x-a) \\
&\quad + \frac{1}{2}(D_h^2 f)(a)(x-a)_h^2 - \frac{1}{6}\int_a^x D_h^3 f(t)d_h(x-t)_h^3
\end{aligned}
$$

and so on, and hence for any nonnegative integer n,

$$
f(x) = \sum_{j=0}^{n} \frac{(D_h^j f)(a)}{j!}(x-a)_h^j - \frac{1}{(n+1)!}\int_a^x D_h^{n+1} f(t)d_h(x-t)_h^{n+1}, \quad (22.17)
$$

or

$$
f(x) = \sum_{j=0}^{n} \frac{(D_h^j f)(a)}{j!}(x-a)_h^j + \frac{1}{n!}\int_a^x (x-t-h)_h^n D_h^{n+1} f(t)d_h t. \quad (22.18)
$$

This formula is called the *Newton interpolation formula*. As we have seen, from the point of view of h-calculus, this is the h-Taylor formula with a remainder term. Suppose $h > 0$. By (22.13), the absolute value of the remainder term is bounded by

$$
\frac{1}{n!}|x-a|^{n+1} \max_{[a,x]} \left| D_h^{n+1} f \right|. \quad (22.19)
$$

To see why (22.18) is called an interpolation formula, let $x = a + mh$, where m is a positive integer. The integral in (22.18) then equals

$$
\sum_{j=0}^{m-1} \left((a+mh)-(a+jh)-h\right)_h^n D_h^{n+1} f(a+jh)
$$

by (22.13). Since the function $g(t) = (a+mh-t-h)_h^n$ vanishes when $t = a+(m-1)h, a+(m-2)h, \ldots, a+(m-n)h$, the remainder term vanishes when m is an integer between 1 and n. The integral obviously also vanishes when $m = 0$. This shows that the finite sum in (22.18) is exactly equal to $f(x)$ at the $n+1$ equally spaced points $x = a, a+h, a+2h, \ldots, a+nh$. Therefore, the sum, when considered as a function of x, is in fact the interpolation polynomial of degree n that approximates an arbitrary function $f(x)$ in the interval $[a, b = a + nh]$. The error term may be estimated using (22.19).

Because of the resemblance between (22.18) and (20.4), a similar discussion is also valid for the q-version, i.e., the sum on the right-hand side of (20.4) may be looked upon as an interpolation polynomial, which is exact at $a, qa, \ldots, q^n a$.

23

Bernoulli Polynomials and Bernoulli Numbers

In this chapter, we introduce a sequence of polynomials that is closely related to the h-antiderivative of polynomials and has many important applications.

Definition. In the Taylor expansion

$$\sum_{n=0}^{\infty} \frac{B_n(x)}{n!} z^n = \frac{ze^{zx}}{e^z - 1}, \qquad (23.1)$$

$B_n(x)$ are polynomials in x, for each nonnegative integer n. They are known as *Bernoulli polynomials*.

If we differentiate both sides of (23.1) with respect to x, we get

$$\sum_{n=0}^{\infty} \frac{B_n'(x)}{n!} z^n = z \frac{ze^{zx}}{e^z - 1} = \sum_{n=0}^{\infty} \frac{B_n(x)}{n!} z^{n+1}.$$

Equating coefficients of z^n, where $n \geq 1$, yields

$$B_n'(x) = nB_{n-1}(x). \qquad (23.2)$$

Together with the fact that $B_0(x) = 1$, which may be obtained by letting z tend to zero on both sides of (23.1), it follows that the degree of $B_n(x)$ is n and its leading coefficient is unity. Using (23.2), we can determine $B_n(x)$ one by one, provided that their constant terms are known.

Definition. For $n \geq 0$, $b_n = B_n(0)$ are called *Bernoulli numbers*.

Putting $x = 0$ in (23.1), we get

$$\sum_{n=0}^{\infty} \frac{b_n}{n!} z^n = \frac{z}{e^z - 1}. \tag{23.3}$$

Since using Taylor's expansion we have

$$\frac{z}{e^z - 1} = \frac{1}{1 + \frac{z}{2} + \frac{z^2}{6} + \frac{z^3}{24} + \cdots},$$

we may use long division to find the Bernoulli numbers. However, we would like to determine b_n and $B_n(x)$ in an easier and more systematic way. To achieve this, we need the following propositions.

Proposition 23.1. *For any $n \geq 1$,*

$$B_n(x + 1) - B_n(x) = nx^{n-1}. \tag{23.4}$$

Proof. Comparing the coefficient of z^n in

$$\sum_{n=0}^{\infty} \frac{B_n(x+1)}{n!} z^n - \sum_{n=0}^{\infty} \frac{B_n(x)}{n!} z^n = \frac{z e^{z(x+1)} - z e^{zx}}{e^z - 1} = z e^{zx} = \frac{d}{dx} e^{zx},$$

where

$$e^{zx} = \sum_{n=0}^{\infty} \frac{x^n z^n}{n!},$$

we have

$$B_n(x + 1) - B_n(x) = \frac{d}{dx} x^n = nx^{n-1},$$

as desired. □

Proposition 23.2. *For any $n \geq 0$,*

$$B_n(x) = \sum_{j=0}^{n} \binom{n}{j} b_j x^{n-j}. \tag{23.5}$$

Proof. Let

$$F_n(x) = \sum_{j=0}^{n} \binom{n}{j} b_j x^{n-j}.$$

It suffices to show that (i) $F_n(0) = b_n$ for $n \geq 0$ and (ii) $F_n'(x) = nF_{n-1}(x)$ for any $n \geq 1$, since these two properties uniquely characterize $B_n(x)$. The first property is obvious. As for the second property, using the fact that for $n > j \geq 0$,

$$(n - j)\binom{n}{j} = \frac{n!}{j!(n - j - 1)!} = n\binom{n-1}{j},$$

we have for $n \geq 1$,

$$\frac{d}{dx} F_n(x) = \sum_{j=0}^{n-1} \binom{n}{j} (n-j) b_j x^{n-1-j} = n \sum_{j=0}^{n-1} \binom{n-1}{j} b_j x^{n-1-j},$$

as desired.   □

Putting $x = 1$ in (23.5), we have

$$B_n(1) = \sum_{j=0}^{n} \binom{n}{j} b_j = b_n + \sum_{j=0}^{n-1} \binom{n}{j} b_j, \quad n \geq 1.$$

However, for any $n \geq 2$, we have $B_n(1) = b_n$, which follows from (23.4) with $x = 0$. Therefore, we obtain the formula

$$\sum_{j=0}^{n-1} \binom{n}{j} b_j = 0, \quad n \geq 2. \tag{23.6}$$

This formula allows us to compute the Bernoulli numbers inductively. The first few of them are

$$b_0 = 1, \ b_1 = -\frac{1}{2}, \ b_2 = \frac{1}{6}, \ b_3 = 0, \ b_4 = -\frac{1}{30}, \ b_5 = 0, \ b_6 = \frac{1}{42}. \tag{23.7}$$

It is tempting to guess that $|b_n| \to 0$ as $n \to \infty$. However, if we consider some other numbers in the sequence,

$$b_8 = -\frac{1}{30}, \ b_{10} = \frac{5}{66}, \ b_{12} = -\frac{691}{2730}, \ b_{14} = \frac{7}{6},$$

$$b_{16} = -\frac{3617}{510}, \ b_{18} = \frac{43867}{798}, \ b_{20} = -\frac{174611}{330},$$

we notice that their values are in general growing with alternating sign (see a discussion of this question in Chapter 25). Another important property of the Bernoulli numbers is $b_n = 0$ for odd $n \geq 3$, which follows from the fact that the function

$$f(z) = \sum_{n=0}^{\infty} \frac{b_n}{n!} z^n - b_1 z = \frac{z}{e^z - 1} + \frac{z}{2} = \frac{z}{2} \frac{e^z + 1}{e^z - 1}$$

is even, i.e., $f(-z) = f(z)$. (The coefficients of t^n in the Taylor expansion about 0 of any even function $g(t)$ vanish for all odd n, because if g is even, $g^{(n)}(t) = (-1)^n g^{(n)}(-t)$ for any n and $g^{(n)}(0) = -g^{(n)}(0)$ for any odd n.)

Another interesting formula involving the Bernoulli numbers may be obtained by putting $x = -1$ into both (23.4) and (23.5), which yields

$$b_n + n(-1)^n = B_n(-1) = \sum_{j=0}^{n} \binom{n}{j} b_j (-1)^{n-j},$$

or

$$n = 1 + \frac{n}{2} + \sum_{j=2}^{n-1} (-1)^j \binom{n}{j} b_j.$$

Replacing j by $j + 1$, and n by $n + 1$, we get:

$$\sum_{j=1}^{n-1} (-1)^j \binom{n}{j} \frac{b_{j+1}}{j+1} = \frac{1}{n+1} - \frac{1}{2}. \tag{23.8}$$

Proposition 23.3. *For any $n \geq 1$,*

$$\sum_{j=0}^{n-1} \binom{n}{j} B_j(x) = n x^{n-1}. \tag{23.9}$$

Proof. The case where $n = 1$ is obvious. If we assume that (23.9) is true for some $k \geq 1$, we have, by (23.2),

$$
\begin{aligned}
\frac{d}{dx} \sum_{j=0}^{k} \binom{k+1}{j} B_j(x) &= \sum_{j=1}^{k} j \binom{k+1}{j} B_{j-1}(x) \\
&= (k+1) \sum_{j=1}^{k} \binom{k}{j-1} B_{j-1}(x) \\
&= (k+1) \sum_{j=0}^{k-1} \binom{k}{j} B_j(x) \\
&= (k+1) k x^{k-1} = (k+1) \frac{d}{dx} x^k,
\end{aligned}
$$

or

$$\sum_{j=0}^{k} \binom{k+1}{j} B_j(x) = (k+1) x^k + C,$$

for some constant C. Putting $x = 0$ and using (23.6) show that $C = 0$. Hence, by induction, (23.9) is true for any positive integer. $\square$

As has been mentioned above, formula (23.2) and the knowledge of Bernoulii numbers allow us to determine the Bernoulli polynomials inductively. The first six of them are listed below:

$$
\begin{aligned}
B_0(x) &= 1, \\
B_1(x) &= x - \frac{1}{2}, \\
B_2(x) &= x^2 - x + \frac{1}{6}, \\
B_3(x) &= x^3 - \frac{3}{2}x^2 + \frac{1}{2}x, \\
B_4(x) &= x^4 - 2x^3 + x^2 - \frac{1}{30}, \\
B_5(x) &= x^5 - \frac{5}{2}x^4 + \frac{5}{3}x^3 - \frac{1}{6}x.
\end{aligned}
$$

24
Sums of Powers

We now turn to the relation between the Bernoulli polynomials and h-calculus. By Proposition 23.1, we have

$$D_1 B_n(x) = B_n(x+1) - B_n(x) = nx^{n-1},$$

or

$$n \int x^{n-1} d_1 x = B_n(x), \tag{24.1}$$

where D_1 is the h-derivative with $h = 1$ and $\int f(x) d_1 x$ stands for the h-antiderivative with $h=1$. Applying the fundamental theorem of h-calculus (22.14), we have for a nonnegative integer n,

$$a^n + (a+1)^n + \cdots + (b-1)^n = \int_a^b x^n d_1 x = \frac{B_{n+1}(b) - B_{n+1}(a)}{n+1}, \tag{24.2}$$

where $a < b$ and $b - a \in \mathbb{Z}$. If we rewrite the right-hand side using (23.5) and let $a = 0$, $b = M + 1$, we get

$$\sum_{k=0}^{M} k^n = \frac{1}{n+1} \sum_{j=0}^{n} \binom{n+1}{j} (M+1)^{n+1-j} b_j. \tag{24.3}$$

Once the Bernoulli numbers are known, one can use (24.3) to easily find the formulas for summing integer powers.

For example, when $n = 2$, (24.3) becomes

$$\sum_{k=1}^{M} k^2 = \frac{1}{3}\left((M+1)^3 - \frac{3}{2}(M+1)^2 + \frac{1}{2}(M+1)\right) = \frac{M(M+1)(2M+1)}{6}.$$

When $n = 3$, we have

$$\sum_{k=1}^{M} k^3 = \frac{1}{4}\left((M+1)^4 - 2(M+1)^3 + (M+1)^2\right) = \left(\frac{M(M+1)}{2}\right)^2.$$

It is an interesting coincidence that for any positive integer M,

$$1^3 + 2^3 + \cdots + M^3 = (1 + 2 + \cdots + M)^2.$$

Also, (24.3) reveals the general fact that the formula for $1^n + 2^n + \cdots + M^n$ is a polynomial in M of degree $n + 1$. This polynomial, for any positive integer n, contains the factor $M(M+1)$, since the right-hand side of (24.3) obviously vanishes at $M = -1$ and, by (23.6), also vanishes at $M = 0$.

25
Euler–Maclaurin Formula

In q-calculus, the Jackson formula (19.2) provides a way to compute explicitly a q-antiderivative of any function. Recall that the Jackson formula was deduced formally using operators. We will do a similar thing for the h-antiderivative in this chapter.

Suppose $D_h F(x) = f(x)$. Using the ordinary Taylor formula, we have

$$F(x + h) = \sum_{n=0}^{\infty} \frac{F^{(n)}(x)h^n}{n!} = \left(\sum_{n=0}^{\infty} \frac{h^n D^n}{n!} \right) F(x),$$

and hence, formally,

$$F(x + h) = e^{hD} F(x), \tag{25.1}$$

where $D \equiv \frac{d}{dx}$. Thus, we have

$$f(x) = \frac{F(x + h) - F(x)}{h} = \frac{e^{hD} - 1}{h} F(x),$$

or

$$F(x) = \frac{hD}{e^{hD} - 1} \int f(x)dx. \tag{25.2}$$

By the definition of Bernoulli numbers (23.3), we have

$$\frac{hD}{e^{hD} - 1} = \sum_{n=0}^{\infty} \frac{b_n}{n!} (hD)^n,$$

and hence

$$F(x) = \sum_{n=0}^{\infty} \frac{b_n}{n!}(hD)^n \int f(x)dx.$$

Using the fact that $b_n = 0$ if n is odd and $n \geq 3$, we deduce the *Euler–Maclaurin formula*

$$F(x) = \int f(x)dx - \frac{h}{2}f(x) + \sum_{n=1}^{\infty} \frac{b_{2n}h^{2n}}{(2n)!}f^{(2n-1)}(x). \qquad (25.3)$$

Note that the ordinary integral and derivative are involved in (25.3). Suppose $h = 1$ and $b - a \in \mathbb{N}$. Using the fundamental theorem of h-calculus, we have

$$\sum_{n=a}^{b-1} f(n) = \int_a^b f(x)dx - \frac{1}{2}\Big(f(b) - f(a)\Big)$$

$$+ \sum_{n=1}^{\infty} \frac{b_{2n}}{(2n)!}\Big(f^{(2n-1)}(b) - f^{(2n-1)}(a)\Big). \qquad (25.4)$$

If f decreases so rapidly with x that all its derivatives approach zero as $x \to \infty$, we have

$$\sum_{n=a}^{\infty} f(n) = \int_a^{\infty} f(x)dx + \frac{1}{2}f(a) - \sum_{n=1}^{\infty} \frac{b_{2n}}{(2n)!}f^{(2n-1)}(a). \qquad (25.5)$$

In order to justisfy the formal derivation of these formulas, let us consider some examples. If $f(x) = x^s$, where s is a positive integer, (25.4) becomes

$$\sum_{n=a}^{b-1} n^s = \frac{b^{s+1} - a^{s+1}}{s+1} - \frac{b^s - a^s}{2}$$

$$+ \sum_{m=2}^{s+1} \frac{b_m}{m!} s(s-1)\cdots(s - m + 2)(b^{s-m+1} - a^{s-m+1})$$

$$= \frac{1}{s+1}\left(x^{s+1} - \frac{s+1}{2}x^s + \sum_{m=2}^{s+1} \binom{s+1}{m} b_m x^{s+1-m}\right)\Bigg|_{x=a}^{x=b},$$

which, by (23.5), is simply $\frac{1}{s+1}(B_{s+1}(b) - B_{s+1}(a))$. We have just recovered formula (24.2) from the previous chapter. As a second example, consider (25.5) with $f(x) = e^{-x}$ and $a = 0$. The LHS is a geometric series,

$$\sum_{n=0}^{\infty} (e^{-1})^n = \frac{1}{1 - e^{-1}},$$

which agrees with the RHS,

$$1 + \frac{1}{2} + \sum_{n=1}^{\infty} \frac{b_{2n}}{(2n)!} = \frac{-1}{e^{-1} - 1},$$

where we have used (23.3) with $z = -1$. Hence, the Euler–Maclaurin formula has more than merely formal value, at least for functions decreasing rapidly at infinity, like $f(x) = e^{-x}$.

For formulas (25.4) and (25.5), although both their left- and right-hand sides involve summations, those on the RHS *may* converge much faster than those on the LHS. In that case, the formulas provide efficient ways to estimate finite and infinite sums. However, we must be careful in applying these formulas to estimate a sum, because, as discussed in Chapter 23, $|b_{2n}|$ increases indefinitely with n. Consider $f(x) = x^{-2}$. Since $f^{(n)}(x) = (-1)^n 2 \cdot 3 \cdots (n+1) x^{-n-2}$, we have

$$\sum_{n=a}^{\infty} \frac{1}{n^2} = \frac{1}{a} + \frac{1}{2a^2} + \sum_{n=1}^{\infty} \frac{b_{2n}}{a^{2n+1}}. \tag{25.6}$$

We will see that the series on the RHS first converges rapidly, but at some sufficiently large n, $|b_{2n}|$ becomes dominant over a^{2n}, and the partial sum bounces up and down more and more drastically. In order to see how well a certain partial sum approximates the actual value, we would like to derive a formula similar to (25.4), but with the infinite sum on the RHS replaced by the Nth partial sum, s_N, plus an additional remainder term, R_N.

To begin with, we rewrite the RHS of (25.4) as

$$\sum_{k=0}^{\infty} \frac{b_k}{k!} \left(h^{(k)}(b) - h^{(k)}(a) \right),$$

where $h(x) = \int f(x)dx$. Suppose $a \in \mathbb{Z}$ and $b = a + 1$. Consider the Nth partial sum,

$$s_N = \sum_{k=0}^{N} \frac{B_k(0)}{k!} \left(h^{(k)}(a+1) - h^{(k)}(a) \right),$$

where N is a positive integer. If we let $g(x) = B_N(x)/N!$, we have $g^{(N-k)}(x) = B_k(x)/k!$ by (23.2). By Proposition 23.1, we have $B_k(1) = B_k(0)$ for $k \neq 1$ and $B_1(1) = B_1(0) + 1$. Hence

$$s_N = \sum_{k=0}^{N} \left(\frac{B_k(0)}{k!} h^{(k)}(a+1) - \frac{B_k(1)}{k!} h^{(k)}(a) \right) + h'(a)$$

$$= \sum_{k=0}^{N} \left(g^{(N-k)}(0) h^{(k)}(a+1) - g^{(N-k)}(1) h^{(k)}(a) \right) + h'(a)$$

$$= h'(a) - \sum_{k=0}^{N} g^{(N-k)}(x) h^{(k)}(a+1-x) \Big|_{x=0}^{x=1}.$$

Since the function

$$G(x) = \sum_{k=0}^{N} g^{(N-k)}(x) h^{(k)}(a+1-x)$$

has a very simple derivative,

$$
\begin{aligned}
G'(x) &= \sum_{k=0}^{N} g^{(N-k+1)}(x) h^{(k)}(a+1-x) \\
&\quad - \sum_{k=0}^{N} g^{(N-k)}(x) h^{(k+1)}(a+1-x) \\
&= \sum_{k=0}^{N} g^{(N-k+1)}(x) h^{(k)}(a+1-x) \\
&\quad - \sum_{k=1}^{N+1} g^{(N-k+1)}(x) h^{(k)}(a+1-x) \\
&= g^{(N+1)}(x) h(a+1-x) - g(x) h^{(N+1)}(a+1-x) \\
&= -g(x) h^{(N+1)}(a+1-x),
\end{aligned}
$$

where $g^{(N+1)}(x) = 0$ because $\deg g = \deg B_N = N$, we have

$$s_N = h'(a) - G(1) + G(0) = h'(a) + \int_0^1 g(x) h^{(N+1)}(a+1-x)dx,$$

or

$$\sum_{k=0}^{N} \frac{b_k}{k!} \left(f^{(k-1)}(a+1) - f^{(k-1)}(a) \right) = f(a) + \int_0^1 \frac{B_N(x)}{N!} f^{(N)}(a+1-x)dx,$$

if we denote $\int f(x)dx$ by $f^{(-1)}(x)$. Performing the change of variable $x = a+1-t$ in the integral, we obtain

$$
\begin{aligned}
f(a) &= \sum_{k=0}^{N} \frac{b_k}{k!} \left(f^{(k-1)}(a+1) - f^{(k-1)}(a) \right) \\
&\quad - \int_a^{a+1} \frac{B_N(\{1-t\})}{N!} f^{(N)}(t)dt,
\end{aligned}
$$

where $\{y\} \in [0,1)$ denotes the fractional part of a real number y. In fact, since a is an integer, for $a < t < a+1$ we have $-a < 1-t < -a+1$ and thus $\{1-t\} = a+1-t$. Finally, replacing a by $a+1, a+2, \ldots, b-1$ and summing all of them yields

$$\sum_{n=a}^{b-1} f(n) = \sum_{k=0}^{N} \frac{b_k}{k!} \left(f^{(k-1)}(b) - f^{(k-1)}(a) \right) - \int_a^b \frac{B_N(\{1-t\})}{N!} f^{(N)}(t)dt,$$

$$(25.7)$$

or, with $N = 2m + 1$,

$$\sum_{n=a}^{b-1} f(n) = \int_a^b f(t)dt - \frac{1}{2}\left(f(b) - f(a)\right)$$

$$+ \sum_{k=1}^{m} \frac{b_{2k}}{(2k)!}\left(f^{(2k-1)}(b) - f^{(2k-1)}(a)\right)$$

$$- \int_a^b \frac{B_{2m+1}(\{1 - t\})}{(2m+1)!} f^{(2m+1)}(t)dt, \qquad (25.8)$$

which is the *Euler–Maclaurin formula with remainder*. If f and its derivatives vanish at infinity, we have

$$\sum_{n=a}^{\infty} f(n) = \int_a^{\infty} f(t)dt + \frac{1}{2}f(a) - \sum_{k=1}^{m} \frac{b_{2k}}{(2k)!} f^{(2k-1)}(a)$$

$$- \int_a^{\infty} \frac{B_{2m+1}(\{1 - t\})}{(2m+1)!} f^{(2m+1)}(t)dt. \qquad (25.9)$$

It follows from (25.7) that we may replace $2m + 1$ by $2m$ in the remainder of (25.8) and (25.9) if $m \geq 1$.

Formulas (25.8) and (25.9) tell us that the error in approximating the sums on the LHS by s_{2m+1} is given by the remainder term,

$$R_{2m+1} = \int_a^b \frac{B_{2m+1}(\{1 - t\})}{(2m+1)!} f^{(2m+1)}(t)dt \quad (b \leq \infty). \qquad (25.10)$$

To estimate its size, we need an upper bound of $B_{2m+1}(x)$ on the interval $[0, 1]$. By the definition of Bernoulli polynomials (23.1), we have

$$\sum_{n=0}^{\infty} \frac{B_n(a)}{n!} z^n = \frac{ze^{az}}{e^z - 1}, \qquad (25.11)$$

for any $0 \leq a \leq 1$. It is known from complex analysis that the radius of convergence of the power series of a function $f(z)$ about $z = 0$ is given by the distance on the complex plane from the origin to the nearest point where $f(z)$ blows up. (For example, it is well known that the radius of convergence of the geometric series $\sum_{n \geq 0} z^n$ is 1, which also follows from the fact that the sum equals $(1 - z)^{-1}$ wherever it converges.) Now, all the points where $(e^z - 1)^{-1}$ blows up are $2\pi ni$, where n is an integer; hence the nearest to the origin among such points are $\pm 2\pi i$. The radius of convergence of the power series in (25.11) is thus 2π.

On the other hand, another fact from analysis tells us that if R is the radius of convergence of the power series $\sum a_n x^n$, then the upper limit of the values $(|a_n|R^n)^{1/n} = R|a_n|^{1/n}$ is unity, i.e., $|a_n|^{1/n}$ is eventually bounded by $1/R$, but by no smaller numbers. (One may justify this fact by considering the geometric series $\sum \alpha^n x^n$, where $R = 1/|\alpha|$.) What all this means is that as n tends to ∞, we have

$$\left(\frac{|B_n(a)|}{n!}\right)^{\frac{1}{n}} \sim \frac{1}{2\pi},$$

or

$$\frac{|B_n(a)|}{n!} \sim \frac{1}{(2\pi)^n}.$$

Hence, following from (25.10), the required bound of $|R_{2m+1}|$ has a size similar to

$$\frac{1}{(2\pi)^{2m+1}} \int_a^b \left|f^{(2m+1)}(t)\right| dt.$$

It is important to note that our discussion of the size of $|R_{2m+1}|$ here is not a rigorous one. However, the above result is valid, since in fact, it is known that for $0 \leq a \leq 1$,

$$\frac{|B_{2m+1}(a)|}{(2m+1)!} < \frac{4e^{2\pi}}{(2\pi)^{2m+1}}. \tag{25.12}$$

(This estimate can be obtained by decomposing $B_{2m+1}(x)$ in a Fourier series.) Therefore, we have

$$|R_{2m+1}| < \frac{4e^{2\pi}}{(2\pi)^{2m+1}} \int_a^b |f^{(2m+1)}(t)|dt. \tag{25.13}$$

For example, with $f(x) = e^{-x}$, $a = 0$, and $b = \infty$, (25.13) tells us that $|R_{2m+1}| < 4e^{2\pi}/(2\pi)^{2m+1}$ goes to zero very rapidly, and we saw earlier in this chapter that the Euler–Maclaurin series does converge.

Next, let us consider $f(x) = x^{-2}$, $b = \infty$ again. Suppose we want to estimate the sum

$$\sum_{n=1}^{\infty} \frac{1}{n^2}, \tag{25.14}$$

and we first do it by adding up a large number of terms, say 1000. The tail of the sum may be estimated using integrals, namely,

$$0.001 = \int_{1000}^{\infty} \frac{dx}{x^2} < \sum_{n=1000}^{\infty} \frac{1}{n^2} < \int_{999}^{\infty} \frac{dx}{x^2} = 0.001001\ldots.$$

We see that our first method gives an answer accurate only to about the sixth decimal place. What if we go further and use (25.9) to approximate the tail? From (25.13), we have the inequality

$$|R_{2m+1}| < \frac{4e^{2\pi}(2m+1)!}{1000(2000\pi)^{2m+1}}.$$

In particular, we have

$$|R_3| < 10^{-10}, \qquad |R_5| < 10^{-16}, \qquad |R_7| < 10^{-23}.$$

This means that a few more calculations will enormously increase the accuracy of our approximations.

It is tempting to conclude that the more terms we compute, the more accurate an estimate we obtain. However, the ratio

$$\left| \frac{R_{2m+1}}{R_{2m-1}} \right| \sim \frac{2m(2m+1)}{(2000\pi)^2}$$

eventually exceeds unity. When m is small, $|R_{2m+1}|$ diminishes rapidly and the partial sums seem to converge, until when $m \approx 1000\pi$, the value of $|R_{2m+1}|$ is minimized, and beyond that, the partial sums start to bounce more and more vigorously away from the actual value. In general, if the tail starts from $n = a$, $|R_{2m+1}|$ is minimized when $m \approx \pi a$.

The exact value of the sum (25.14) was first discovered by Euler to be $\pi^2/6$. There are many ways to prove it, one of which makes use of complex analysis and another involves Fourier series. Both of them also allow us to compute the exact values of $\sum_{n=1}^{\infty}(1/n^s)$, where s is an even positive integer: they turn out to be $2^{s-1}\pi^s|b_s|/s!$. However, no similar closed form has been discovered for these sums when s is odd.

In conclusion consdier two other simple examples of the Euler–MacLaurin formula (25.8), with $a = 1$, $m = 1$.

Example 1. $f(x) = \frac{1}{x}$. Then we have:

$$\sum_{n=1}^{b} \frac{1}{n} = \log b + R(b), \quad \text{where } \lim_{b \to \infty} R(b) = c. \qquad (25.15)$$

The number c is called Euler's constant.

Example 2. $f(x) = \log x$. Then we have:

$$\log(b-1)! = \int_{1}^{b} \log t \, dt - \frac{1}{2}\log b + R_1(b), \quad \text{where } \lim_{b \to \infty} R_1(b) = C. \quad (25.16)$$

Integrating by parts, we get

$$\int_{1}^{b} \log t \, dt = b(\log b - 1).$$

Hence, adding $\log b$ to both sides of (25.16), we get:

$$\log b! = \log \sqrt{b} \left(\frac{b}{e} \right)^{b} + R_1(b).$$

Therefore,

$$b! \sim e^{C} \sqrt{b} \left(\frac{b}{e} \right)^{b} \quad \text{as } b \to \infty. \qquad (25.17)$$

Additional calculations show that $e^{C} = \sqrt{2\pi}$. The result is the famous Stirling's formula.

26
Symmetric Quantum Calculus

The q- and h-differentials may be "symmetrized" in the following way,

$$\tilde{d}_q f(x) = f(qx) - f(q^{-1}x), \tag{26.1}$$
$$\tilde{d}_h g(x) = g(x+h) - g(x-h), \tag{26.2}$$

where as usual, $q \neq 1$ and $h \neq 0$. The definitions of the corresponding derivatives follow obviously:

$$\tilde{D}_q f(x) = \frac{\tilde{d}_q f(x)}{\tilde{d}_q x} = \frac{f(qx) - f(q^{-1}x)}{(q - q^{-1})x}, \tag{26.3}$$

$$\tilde{d}_h g(x) = \frac{\tilde{d}_h g(x)}{\tilde{d}_h x} = \frac{g(x+h) - g(x-h)}{2h}. \tag{26.4}$$

We are going to concern ourselves briefly with symmetric q-calculus only, since it is important for the theory of some algebraic objects called quantum groups.

The symmetric q-product and quotient rules are

$$\tilde{D}_q\big(f(x)g(x)\big) = f(qx)\tilde{D}_q g(x) + g(q^{-1}x)\tilde{D}_q f(x)$$
$$= f(q^{-1}x)\tilde{D}_q g(x) + g(qx)\tilde{D}_q f(x), \tag{26.5}$$

$$\tilde{D}_q\left(\frac{f(x)}{g(x)}\right) = \frac{g(qx)\tilde{D}_q f(x) - f(qx)\tilde{D}_q g(x)}{g(qx)g(q^{-1}x)}$$
$$= \frac{g(q^{-1}x)\tilde{D}_q f(x) - f(q^{-1}x)\tilde{D}_q g(x)}{g(qx)g(q^{-1}x)}. \tag{26.6}$$

For any number α, we have

$$\tilde{D}_q x^\alpha = [\alpha]\tilde{\ } x^{\alpha-1}, \tag{26.7}$$

where

$$[\alpha]\tilde{\ } = \frac{q^\alpha - q^{-\alpha}}{q - q^{-1}}. \tag{26.8}$$

Proposition 26.1. *For any positive integer n, if we define*

$$(x - a)_{\tilde{q}}^n = (x - q^{n-1}a)(x - q^{n-3}a)(x - q^{n-5}a)\cdots(x - q^{-n+1}a), \tag{26.9}$$

and $(x - a)_{\tilde{q}}^0 = 1$, we have

$$\tilde{D}_q(x - a)_{\tilde{q}}^n = [n]\tilde{\ }(x - a)_{\tilde{q}}^{n-1}. \tag{26.10}$$

Proof   The case of $n = 1$ is trivial. Note that for any $n > 1$, $(x-a)_{\tilde{q}}^{n+1} = (x - qa)_{\tilde{q}}^n(x - q^{-n}a)$. Thus, by (26.5) and induction on n, we have

$$
\begin{aligned}
\tilde{D}_q(x - a)_{\tilde{q}}^{n+1} &= (qx - qa)_{\tilde{q}}^n + [n]\tilde{\ }(x - qa)_{\tilde{q}}^{n-1}(q^{-1}x - q^{-n}a) \\
&= q^n(x - a)_{\tilde{q}}^n + q^{-1}[n]\tilde{\ }(x - qa)_{\tilde{q}}^{n-1}(x - q^{-n+1}a) \\
&= (q^n + q^{-1}[n]\tilde{\ })(x - a)_{\tilde{q}}^n = [n + 1]\tilde{\ }(x - a)_{\tilde{q}}^n,
\end{aligned}
$$

as desired. $\square$

Note that $\deg(x - a)_{\tilde{q}}^n = n$ for each n, and the first three of them are

$$
\begin{aligned}
(x - a)_{\tilde{q}}^1 &= (x - a), \\
(x - a)_{\tilde{q}}^2 &= (x - qa)(x - q^{-1}a), \\
(x - a)_{\tilde{q}}^3 &= (x - q^2a)(x - a)(x - q^{-2}a).
\end{aligned}
$$

However, if $a \neq 0$, $(x - a)_{\tilde{q}}^n$ does not vanish at $x = a$ when n is even, and thus the polynomials $P_n(x) = (x-a)_{\tilde{q}}^n/[n]\tilde{\ }!$ do not satisfy all the conditions for the generalized Taylor formula (Theorem 2.1). (We refer to (26.23) for the family of polynomials to which Theorem 2.1 does apply.) For $a = 0$, the Taylor expansion of a formal power series is

$$f(x) = \sum_{j=0}^\infty \left(\tilde{D}_q^j f\right)(0)\frac{x^j}{[j]\tilde{\ }!}. \tag{26.11}$$

Consider $f(x) = (x + a)_{\tilde{q}}^n$. Since $(\tilde{D}_q^j f)(0) = [n]\tilde{\ }[n - 1]\tilde{\ }\cdots[n - j + 1]\tilde{\ } \times (0+a)_q^{n-j} = ([n]\tilde{\ }!/[n-j]\tilde{\ }!)a^{n-j}$ for $j \leq n$ and $\tilde{D}_q^j f(x) = 0$ for $j > n$, we have

$$(x + a)_{\tilde{q}}^n = \sum_{j=0}^n \left[\begin{array}{c} n \\ j \end{array} \right]\tilde{\ } a^{n-j}x^j, \tag{26.12}$$

where

$$\left[\begin{array}{c} n \\ j \end{array} \right]\tilde{\ } = \frac{[n]\tilde{\ }!}{[j]\tilde{\ }![n - j]\tilde{\ }!}. \tag{26.13}$$

Equation (26.12) is the $\tilde{q}$-analogue of Gauss's binomial formula (5.5). We would also like to obtain a $\tilde{q}$-analogue of Heine's binomial formula (8.1). Let us consider $g(x) = 1/(1-x)_{\tilde{q}}^n$. Since

$$
\begin{aligned}
(1-x)_{\tilde{q}}^n &= (1-q^{n-1}x)(1-q^{n-3}x)\cdots(1-q^{1-n}x) \\
&= q^{n-1}(q^{1-n}-x)\cdot q^{n-3}(q^{3-n}-x)\cdots q^{1-n}(q^{n-1}-x),
\end{aligned}
$$

or

$$
(1-x)_{\tilde{q}}^n = (-1)^n(x-1)_{\tilde{q}}^n, \tag{26.14}
$$

we have

$$
\tilde{D}_q(1-x)_{\tilde{q}}^n = (-1)^n[n]^\sim(x-1)_{\tilde{q}}^{n-1} = -[n]^\sim(1-x)_{\tilde{q}}^{n-1},
$$

and, by (26.6),

$$
\tilde{D}_q g(x) = \frac{[n]^\sim(1-x)_{\tilde{q}}^{n-1}}{(1-qx)_{\tilde{q}}^n(1-q^{-1}x)_{\tilde{q}}^n} = \frac{[n]^\sim}{(1-x)_{\tilde{q}}^{n+1}}. \tag{26.15}
$$

Therefore, for any $j \geq 0$, we have $(\tilde{D}_q^j g)(0) = [n]^\sim \cdots [n+j-1]^\sim$, and

$$
\frac{1}{(1-x)_{\tilde{q}}^n} = \sum_{j=0}^{\infty} \frac{[n]^\sim \cdots [n+j-1]^\sim}{[j]^\sim!} x^j, \tag{26.16}
$$

which is very similar to (8.1).

Let us now turn to integration. To derive an explicit formula for the $\tilde{q}$-antiderivative of an arbitrary function $f(x)$, we may again employ a formal approach using operators. Suppose $F(x)$ is a $\tilde{q}$-antiderivative of $f(x)$. Using the operator $\hat{M}_q$ as defined in (5.6), we have

$$
(\hat{M}_q - \hat{M}_{q^{-1}})F(x) = F(qx) - F(q^{-1}x) = (q - q^{-1})xf(x).
$$

Since

$$
\hat{M}_q\hat{M}_{q^{-1}}g(x) = \hat{M}_{q^{-1}}\hat{M}_q g(x) = g(x)
$$

for any $g(x)$, it is natural to write $\hat{M}_{q^{-1}} = (\hat{M}_q)^{-1}$. Thus, we have

$$
\left(\hat{M}_q - \frac{1}{\hat{M}_q}\right)F(x) = (q - q^{-1})xf(x),
$$

or

$$
\begin{aligned}
F(x) &= \frac{\hat{M}_q}{1 - \hat{M}_q^2}(q^{-1} - q)xf(x) \\
&= (q^{-1} - q)(\hat{M}_q + \hat{M}_q^3 + \hat{M}_q^5 + \cdots)xf(x).
\end{aligned}
$$

Hence, we have

$$
F(x) = x(q^{-1} - q) \sum_{n=1,3,\dots} q^n f(q^n x). \tag{26.17}
$$

It is straightforward to verify that if the RHS of (26.17) converges, it does give a $\tilde{q}$-antiderivative of $f(x)$, and this antiderivative vanishes at $x = 0$. Consequently, provided that the series converges, the definite $\tilde{q}$-integral is given by

$$\int_0^a f(x)\tilde{d}_q x = a(q^{-1} - q) \sum_{n=1,3,...} q^n f(q^n a). \tag{26.18}$$

The uniqueness problem rests upon the nature of the solutions to the functional equation $\tilde{D}_q G(x) = 0$. This equation implies $G(qx) = G(q^{-1}x)$, or $G(x) = G(q^{2n}x)$ for any x and integer n. If we require $G(x)$ to be continuous at 0, then $G(x)$ will be arbitrarily close to $G(0)$; thus $G(x)$ is constant. Therefore, as in q-calculus, continuity at $x = 0$ forces the $\tilde{q}$-antiderivative to be uniquely determined up to a constant summand.

If we define

$$\int_a^b f(x)\tilde{d}_q x = \int_0^b f(x)\tilde{d}_q x - \int_0^a f(x)\tilde{d}_q x,$$

we have, in particular,

$$
\begin{aligned}
\int_{q^{m+1}}^{q^{m-1}} f(x)\tilde{d}_q x &= (q^{-1} - q)\left(\sum_{n=1,3,...} q^{n+m-1} f(q^{n+m-1}) \right. \\
&\quad \left. - \sum_{n=1,3,...} q^{n+m+1} f(q^{n+m+1}) \right) \\
&= (q^{-1} - q)q^m f(q^m).
\end{aligned}
$$

It is then natural to define

$$
\begin{aligned}
\int_0^\infty f(x)\tilde{d}_q x &= \sum_{m=\pm 1,\pm 3,...} \int_{q^{m+1}}^{q^{m-1}} f(x)\tilde{d}_q x \\
&= (q^{-1} - q) \sum_{m=\pm 1,\pm 3,...} q^m f(q^m). \tag{26.19}
\end{aligned}
$$

We stop here, leaving it to the reader to develop the symmetric q-calculus further along the lines of the q-calculus.

We conclude the book with a brief discussion of more general quantum calculi. In this book we have encountered three different quantum calculi, namely, the q-calculus, the h-calculus, and the symmetric q-calculus. The most general definition of a quantum differential would be

$$d f(x) = f(qx + h) - f(q'x + h').$$

Similar theories consisting of the corresponding derivatives, Taylor's formulas and antiderivatives may thus be developed. In order that the derivative $Df(x) = d f(x)/dx$, be well-defined, we should assume that either $q \neq q'$ or

$h \neq h'$. A family of polynomials $\{P_n\}_{n \geq 0}$ satisfying all three assumptions of Theorem 2.1 always exists, since such polynomials may be derived one after another starting from $n = 0$. In general, these polynomials have the following expression:

$$P_n(x) = c_n(x - a)(x - a_2) \cdots (x - a_n), \tag{26.20}$$

where a_2, a_3, ... are functions in a, q, q', h, h'. By comparing the leading coefficients in $DP_n(x)$ and $P_{n-1}(x)$, it is easy to see that $c_n/c_{n-1} = 1/[n]$, where we define

$$[n] = (q^n - q'^n)/(q - q') \text{ if } q \neq q' \text{ and } [n] = nq^{n-1} \text{ if } q = q'.$$

Letting $[n]! = [1] \cdots [n]$ for a positive integer n, and $[0]! = 1$, we have

$$c_n = \frac{1}{[n]!}, \quad n \geq 0.$$

However, the general expression of a_n is too complicated to write down explicitly. This time, we compare the coefficients of x^{n-1} in $dP_n(x)$ and $P_{n-1}(x)dx$. This allows us to deduce the following recursion formula for $s_n = a + a_2 + \cdots + a_n$, $n \geq 2$:

$$s_n = \frac{[n]}{[n-1]} s_{n-1}$$
$$+ \left(\frac{n(q^{n-1}h - q'^{n-1}h')}{q^{n-1} - q'^{n-1}} - \frac{[n](h - h')}{q^{n-1} - q'^{n-1}} \right) \quad \text{if } q \neq q', \tag{26.21}$$

$$s_n = \frac{n}{n-1} qs_{n-1} + \frac{1}{2}n(h + h') \quad \text{if } q = q'. \tag{26.22}$$

One can easily see that in general, even a_2 and a_3 have unpleasant expressions. The recursive relation is simpler when $h = h'$. In that case, (26.21) becomes

$$s_n = \frac{[n]}{[n-1]} s_{n-1} + nh, \quad n \geq 2,$$

which, together with the initial condition $s_1 = a$, gives the solution

$$s_n = (a - h)[n] + h[n] \left(\frac{1}{[1]} + \frac{2}{[2]} + \cdots + \frac{n}{[n]} \right), \quad n \geq 1.$$

(The solution may be obtained by means of the subsitution $t_n = s_n/[n]$.) Then, we have

$$
\begin{aligned}
a_n &= s_n - s_{n-1} \\
&= (a - h)([n] - [n-1]) + nh \\
&\quad + h([n] - [n-1]) \left(\frac{1}{[1]} + \cdots + \frac{n-1}{[n-1]} \right), \quad n \geq 2.
\end{aligned}
$$

In particular, for the symmetric q-calculus, i.e., $h = 0$, $q' = q^{-1}$, we have

$$a_n = \left([n]\tilde{} - [n-1]\tilde{}\right)a = (q^{n-1} - q^{n-2} + q^{n-3} - \cdots + q^{1-n})a, \quad n \geq 1.$$

In other words, the sequence of polynomials for symmetric q-calculus to which the generalized Taylor formula applies is given by

$$
\begin{aligned}
P_n(x) \;=\; & \frac{1}{[n]\tilde{}!}(x-a)\left(x - (q - 1 + q^{-1})a\right)\cdots \\
& \times \left(x - (q^{n-1} - q^{n-2} + q^{n-3} - \cdots + q^{1-n})a\right). \quad (26.23)
\end{aligned}
$$

Appendix

Appendix: A List of q-Antiderivatives

$$\int x^{\alpha} d_q x = \frac{x^{\alpha+1}}{[\alpha+1]} \qquad (\alpha \neq -1)$$

$$\int \frac{d_q x}{x} = \frac{q-1}{\log q} \log x$$

$$\int (x-a)_q^{\alpha} d_q x = \frac{(x-a)_q^{\alpha+1}}{[\alpha+1]} \qquad (\alpha \neq -1)$$

$$\int (a-x)_q^{\alpha} d_q x = -\frac{q(a-q^{-1}x)_q^{\alpha+1}}{[\alpha+1]} \qquad (\alpha \neq -1)$$

$$\int \frac{d_q x}{(x-a)_q^{\alpha}} = \frac{1}{q[1-\alpha](x-qa)_q^{\alpha-1}} \qquad (\alpha \neq 1)$$

$$\int \frac{d_q x}{(a-x)_q^{\alpha}} = \frac{1}{[\alpha-1](a-x)_q^{\alpha-1}} \qquad (\alpha \neq 1)$$

$$\int e_q^{\alpha x} d_q x = \frac{1}{\alpha} e_q^{\alpha x}$$

$$\int E_q^{\alpha x} d_q x = \frac{q}{\alpha} e_q^{q^{-1}\alpha x}$$

$$\int \cos_q(\alpha x) d_q x = \frac{1}{\alpha} \sin_q(\alpha x)$$

$$\int \sin_q(\alpha x) d_q x = -\frac{1}{\alpha} \cos_q(\alpha x)$$

$$\int \mathrm{Cos}_q(\alpha x) d_q x = \frac{q}{\alpha} \mathrm{Sin}_q(q^{-1}\alpha x)$$

$$\int \mathrm{Sin}_q(\alpha x) d_q x = -\frac{q}{\alpha} \mathrm{Cos}_q(q^{-1}\alpha x)$$

Integration by parts:

$$\int_a^b f(x)d_qg(x) = f(b)g(b) - f(a)g(a) - \int_a^b g(qx)d_qf(x)$$

Change of variable:

$$\int_{u(a)}^{u(b)} f(u)d_qu = \int_a^b f(u(x))d_{q^{1/\beta}}u(x), \qquad \text{where } u(x) = \alpha x^\beta$$

Literature

G.E. Andrews, *q*-Series: their development and application in analysis, number theory, combinatorics, physics, and computer algebra, *CBMS Regional Conference Lecture Series in Mathematics* **66**, AMS, 1986.

H. Exton, *q-Hypergeometric Functions and Applications*, Halsted Press, Chichister, 1983.

N.J. Fine, *Basic hypergeometric series and applications*, Math. Surveys and Monographs, **27**, AMS, 1988.

G. Gasper and M. Rahman, *Basic Hypergeometric Series*, Cambridge University Press, 1990.

G.H. Hardy and E.M. Wright, *An Introduction to the Theory of Numbers*, Oxford University Press, 1960.

A.O. Gelfond, *Calculus of Finite Differences*, Nauka, Moscow, 1967.

J. Goldman and G-C. Rota, The number of subspaces of a vector space, in *Recent Progress in Combinatorics*, Acad. Press, 1969.

A.A. Kirillov, *Additional Chapters of Mathematical Analysis*, MK IMU, 1994.

H. Miller, *Euler–Maclaurin Formula*, MIT lecture notes, 1998.

N.J. Vilenkin and A.U. Klimyk, *Representations of Lie Groups and Special Functions*, vol.**3**, Kluwer Acad. Publishers, 1992.

Index

Universitext   *(continued)*

Kannan/Krueger: Advanced Analysis
Kelly/Matthews: The Non-Euclidean Hyperbolic Plane
Kostrikin: Introduction to Algebra
Luecking/Rubel: Complex Analysis: A Functional Analysis Approach
MacLane/Moerdijk: Sheaves in Geometry and Logic
Marcus: Number Fields
McCarthy: Introduction to Arithmetical Functions
Meyer: Essential Mathematics for Applied Fields
Mines/Richman/Ruitenburg: A Course in Constructive Algebra
Moise: Introductory Problems Course in Analysis and Topology
Morris: Introduction to Game Theory
Poizat: A Course In Model Theory: An Introduction to Contemporary Mathematical Logic
Polster: A Geometrical Picture Book
Porter/Woods: Extensions and Absolutes of Hausdorff Spaces
Radjavi/Rosenthal: Simultaneous Triangularization
Ramsay/Richtmyer: Introduction to Hyperbolic Geometry
Reisel: Elementary Theory of Metric Spaces
Ribenboim: Classical Theory of Algebraic Numbers
Rickart: Natural Function Algebras
Rotman: Galois Theory
Rubel/Colliander: Entire and Meromorphic Functions
Sagan: Space-Filling Curves
Samelson: Notes on Lie Algebras
Schiff: Normal Families
Shapiro: Composition Operators and Classical Function Theory
Simonnet: Measures and Probability
Smith: Power Series From a Computational Point of View
Smith/Kahanpää/Kekäläinen/Traves: An Invitation to Algebraic Geometry
Smoryski: Self-Reference and Modal Logic
Stillwell: Geometry of Surfaces
Stroock: An Introduction to the Theory of Large Deviations
Sunder: An Invitation to von Neumann Algebras
Tondeur: Foliations on Riemannian Manifolds
Toth: Finite Möbius Groups, Minimal Immersions of Spheres, and Moduli
Wong: Weyl Transforms
Zhang: Matrix Theory: Basic Results and Techniques
Zong: Sphere Packings
Zong: Strange Phenomena in Convex and Discrete Geometry